AF581708

After the Anthropocene

After the Anthropocene: Time and Mobility

Editors

Pasi Heikkurinen
Toni Ruuska
Anu Valtonen
Outi Rantala

MDPI • Basel • Beijing • Wuhan • Barcelona • Belgrade • Manchester • Tokyo • Cluj • Tianjin

Editors
Pasi Heikkurinen
University of Helsinki
Finland

Toni Ruuska
University of Helsinki
Finland

Anu Valtonen
University of Lapland
Finland

Outi Rantala
University of Lapland
Finland

Editorial Office
MDPI
St. Alban-Anlage 66
4052 Basel, Switzerland

This is a reprint of articles from the Special Issue published online in the open access journal *Sustainability* (ISSN 2071-1050) (available at: https://www.mdpi.com/journal/sustainability/special_issues/after_anthropocene_time_mobility).

For citation purposes, cite each article independently as indicated on the article page online and as indicated below:

LastName, A.A.; LastName, B.B.; LastName, C.C. Article Title. *Journal Name* **Year**, *Article Number*, Page Range.

ISBN 978-3-03936-956-0 (Hbk)
ISBN 978-3-03936-957-7 (PDF)

Cover image courtesy of Risto Musta.

Contents

About the Editors

Pasi Heikkurinen Senior lecturer in Management at the University of Helsinki, Department of Economics and Management; visiting lecturer in Business and Sustainable Change at the University of Leeds, Sustainability Research Institute; and docent in Sustainability and Organizations at the Aalto University School of Business. His research project is the phenomenology of sustainability. The focus of his work is on questions concerning ethics and technology in relation to sustainable change, and in particular in the context of food and agriculture.

Toni Ruuska Postdoctoral researcher at the University of Helsinki, Faculty of Agriculture and Forestry, Department of Economics. Toni's research interests are located generally in political ecology and economy. In his post-doctoral research, he is studying the politics of self-sufficiency in food production. The take on the subject is philosophical, i.e., trying to understand what nature is and what kind of knowledge a subject can have of it, but also personal and practical. This is because sustainable change in self-sufficient food production is approached through alternative actions and the practices of individuals and communities by utilising an auto-ethnographic method.

Anu Valtonen Professor of Cultural Economy at the University of Lapland, Faculty of Social Sciences, Finland. She works at the interface of marketing, organization and tourism studies. Her research interests relate to critical and feminist theories, qualitative methodologies, bodies, senses, and sleep cultures. Recently, she has been engaging with feminist new materialism, affect theories, and more-than-human methodologies. Her work has been published in Qualitative Inquiry, Human Relations, Management Learning, Organization, Annals of Tourism Research, Journal of Material Culture, Journal of Marketing Management, Consumption, Markets and Culture, and Tourism Studies, as well as in books and book chapters.

Outi Rantala Associate Professor of Responsible Arctic Tourism at the University of Lapland, in the Multidimensional Tourism Institute/Faculty of Social Sciences. Outi's research focuses on nature relationships and the rhythms of everyday and holidays through such phenomenon as wilderness guiding, sleeping outdoors, weather, adventure and architecture. She engages with reflective ethnographic methodology, post-human practice theory and new materialism, and her research has been published in Annals of Tourism Research, Journal of Material Culture and Scandinavian Journal of Hospitality and Tourism.

Editorial

Time and Mobility after the Anthropocene

Pasi Heikkurinen [1,*], Toni Ruuska [1], Anu Valtonen [2] and Outi Rantala [3]

[1] Department of Economics and Management, University of Helsinki, 00500 Helsinki, Finland; toni.ruuska@helsinki.fi
[2] Faculty of Social Sciences, University of Lapland, 96300 Rovaniemi, Finland; anu.valtonen@ulapland.fi
[3] Multidimensional Tourism Institute, University of Lapland, 96300 Rovaniemi, Finland; outi.rantala@ulapland.fi
* Correspondence: pasi.heikkurinen@helsinki.fi

Received: 29 May 2020; Accepted: 4 June 2020; Published: 24 June 2020

Abstract: The Special Issue on 'After the Anthropocene: Time and Mobility' is published. It discusses the geological time to follow the human-dominated epoch and ways to move there. In addition to this editorial, a total of five articles are published in the issue. The articles engage with a variety of social science disciplines—ranging from economics and sociology to philosophy and political science—and connect to the natural science insights on the Anthropocene. The issue calls for going beyond anthropocentrism in sustainability theory and practice in order to exit the Anthropocene with applications and insights in the contexts of politics (Ruuska et al., 2020), energy (Mohorčich, 2020), tourism (Rantala et al., 2020), food (Mazac and Tuomisto, 2020) and management (Küpers, 2020). We hope that you will find this Special Issue interesting and helpful in contributing to sustainable change.

Keywords: Anthropocene; time; mobility; nature; culture; sustainability

---

1. Introduction

The Anthropocene, characterized by global scale anthropogenic forcing since the latter part of the 18th century, is the name of the present geological epoch [1]. The empirical observations in Earth Sciences about the Anthropocene signify that a single species, *Homo Sapiens*, has grown its impacts on the rest of nature so vast that an exit from the previous epoch, Holocene, is seen as stratigraphically legitimate. The manifestations of the undesired global anthropogenic impacts on Earth include climate change and the loss of biodiversity.

Similarly, as the Holocene ended few centuries ago, it is imaginable that also one day the Anthropocene will be history. In fact, it seems quite obvious that sooner or later, the Earth will reach the end of the Anthropocene. The move to this time will be either with humans or without humans. The normative standpoint we wish to take here is that the Earth should move to this new post-Anthropocene time and humans should be active in this mission. Furthermore, it would be very desirable for humans themselves were they also to inhabit the time to follow the Anthropocene.

The scholarly debate to date has paid relatively little attention to this space–time. Instead, the discussion continues to revolve around questions such as: when did the human-dominated epoch began; what to call it; who or what is to blame for it; and how we might respond to it in the immediate future [2]. While these questions certainly deserve consideration, effort should also be aimed at questions of how the Anthropocene might come to an end (as a discourse and as an epoch); what post-Anthropocene might look like; and what this might signify for organizing social change, and/or caring for the nonhuman nature [3].

As the effects of changing climatic regimes impose greater effects on earthbound habitation and the known ways of being in the present geological epoch, there is a need to consider how humans and/or socio-nature might and should respond. That is, sustainability scholarship should begin imagining a time after the Anthropocene, when humans would no longer be the dominant species on the planet, and think of how to move there. We also consider the notion of the 'late Anthropocene' relevant for discussing the present when humanity—albeit in different place-specific ways—is forced to adapt in radical ways to the challenges that it faces.

2. Peaceful Coexistence Colloquia

The Special Issue 'After the Anthropocene: Time and Mobility' is titled after the theme of "The 3rd Peaceful Coexistence Colloquium" held in Helsinki 13–14 June 2019. This colloquium was organized by the University of Helsinki, Department of Economics and Management together with Sustainable Change Research Network (SUCH). "The 1st Peaceful Coexistence Colloquium" took place in Costa Rica at the University for Peace in 2015 with a theme 'Genders, Natures, and Technologies in the Anthropocene'. The second colloquium was hosted by University of Lapland in Pyhätunturi in 2017 with a theme "Reimagining Ethics and Politics of Space for the Anthropocene". The colloquiums are transdisciplinary meeting places for researchers, activists and artists who challenge the mainstream solutions for the challenges that face the Earth and its habitants. One of the aims of the Peaceful Coexistence Colloquium is also to call for alternative and more radical ways to address the current socio-economic crises.

All of these colloquiums have now produced an edited volume. The book "Sustainability and Peaceful Coexistence for the Anthropocene" connected to the first conference was edited by Pasi Heikkurinen and published in 2017 by Routledge [4]. To analyze the root problems and consequences of unsustainable development, as well as to outline rigorous solutions for the contemporary epoch, this volume brings together natural and social sciences under the rubric of the Anthropocene. The book identifies the central preconditions for social organization and governance to enable the peaceful coexistence of humans and the non-human world. The contributors investigate the burning questions of sustainability from a number of different perspectives including geosciences, economics, law, organizational studies, political theory, and philosophy. The book is a state-of-the-art review of the Anthropocene debate and provides crucial signposts for how human activities can, and should, be changed.

The book "Ethics and Politics of Space for the Anthropocene" linked to the second conference was edited by Anu Valtonen, Outi Rantala and Paolo Davide Farah and will be published in 2020 by Edward Elgar Publishing [5]. The book outlines new and more radical ways to address the current environmental crisis by envisaging a narrative of change that renders visible the range of transformations taking place throughout the globe. This enables the authors to capture the complex nature of ongoing transformations and to alter unjust practices and power structures in a sustainable and context-specific manner. Importantly, the new narrative highlights the localized and situated nature of the Anthropocene, allowing the differences of regions and contexts—and subsequent ethical and political questions—to be taken seriously. It also highlights the potentiality residing in non-Western ways of relating to and living on the earth, taking more-than-humans into account.

This Special Issue "After the Anthropocene: Time and Mobility" is an output from the third conference and published in 2020 by the journal "Sustainability". At the dawn of spring 2019, we were contacted by the Editorial Board of the journal concerning a Special Issue and we suggested the same theme and title as the third Peaceful Coexistence. In the opening words of the colloquium the Special Issue was introduced to the participants, which were also invited to write an article to it. The colloquium itself had 35 participants from 10 different countries. The presentations and talks under the theme were diverse ranging from deep ecology, and indigenous food sovereignty to degrowth, and non-human hauntology. This final edited volume will end the Anthropocene-thematic trilogy in the Peaceful Coexistence collective.

"The 4th Peaceful Coexistence Colloquium" is going to be held in 2021 and organized by Sustainable Change Research Network (SUCH). We invite all interested persons to contact us.

3. Articles in the Issue

This Special Issue comprises five articles, which all explore and discuss ways out of the Anthropocene, as well as envisage possibilities for diverse life after the Anthropocene. The focus of the issue is on questions of time and mobility, insofar as these concepts enrich our understandings of what comes after the Anthropocene and how could an exit from the Anthropocene materialize. The articles explore time and mobility after the Anthropocene in different ways, from a rich diversity theoretical and empirical point of views.

In their article 'Domination, Power, Supremacy: Confronting Anthropolitics with Ecological Realism,' Ruuska et al. [6] studied politics as domination. They claimed that domination, especially in the Anthropocene, has two vital components, which are power and supremacy. Domination does not occur without the power over others. In addition, domination requires reasoning, justification, and legitimation that are often connected to superiority arguments (based on religion, society, civilization, etc.) from the oppressor's end. The authors argued that past and present political ideologies and agendas, such as colonial rule, imperialism, neoliberal capitalism, and also the popular Green New Deal are examples of 'anthropolitics', entailing an anthropocentric approach to politics. In contrast to historical and prevailing anthropolitical programs, the article discusses post-Anthropocene politics under a theoretical frame called ecological realism, which is characterized by equality among beings, and localization and decentralization, as well as steep reduction of matter-energy throughput in the human realm.

In his article 'Energy Intensity and Human Mobility after the Anthropocene,' Mohorčich [7] claimed that after the Anthropocene, human settlements will likely have less available energy to move people and things. His article considered the feasibility of five modes of transportation under two energy-constrained scenarios. Mohorčich analyzed the effects that transformation mode choice is likely to have on the size of post-Anthropocene human settlements, and also what is the role of speed and energy in them. He argued that cars (including battery-electric vehicles) are not feasible under a highly energy-constrained situation, in contrast to buses, metros and walking which are feasible, but limit human settlement size. Cycling is likely the only mode of transport that would enable suburbs in an energy-constrained post-Anthropocene scenario, the article concludes.

In their article 'Envisioning Tourism and Proximity after the Anthropocene', Rantala et al. [8] called for new imaginings, conceptualizations and practices of tourism for the current Earthly crisis. Thus, they conceptualized proximity tourism with feminist new materialist literature, which accords agency to the ongoing common worlding of all matter—including but not limited to humans—rather than to separate individual agents. More specifically, they explore the idea of proximity by drawing closer to the geo—to the Earth—through geological walks in the Pyhä National Park in Finnish Lapland. These walks are analyzed with the notions of rhythmicity, vitality and care—ideas constructed from the theoretical heritage guiding the study. By doing this, Rantala, Salmela, Valtonen, and Höckert explore the potential of proximity tourism in ways that intertwine non-living and living matter, science stories, history, local communities, and tourism. The outcome of this analysis is that they compose one possible narrative of tourism after the Anthropocene.

In their article 'The Post-Anthropocene Diet: Navigating Future Diets for Sustainable Food Systems,' Mazac and Tuomisto [9] examined how future diets could reduce the environmental impacts of food systems, and thus, enable movement into the post-Anthropocene. The authors claimed that non-anthropocentric diets could address global food systems challenges in the Anthropocene. In order to change diets, changes in ontology is proposed. In their article, Mazac and Tuomisto employed indigenous worldviews and object-oriented ecosophy to investigate the possibilities of non-anthropocentric worldviews with a focus on temporality. While indigenous ontologies are introduced as pre-Anthropocene examples that depict humans and non-humans in relational diets, a post-Anthropocene illustration stresses non-dualist object-oriented ecosophy. As a central implication, the article offered ontologically based ideas to developing dietary guidelines for the time after the Anthropocene.

In his article 'From the Anthropocene to an "Ecocene"—Eco-Phenomenological Perspectives on Embodied, Anthrodecentric Transformations towards Enlivening Practices of Organising Sustainably', Küpers [10] discussed the Anthropocene from an eco-phenomenological point of view. The author drew, in particular, on the work of Maurice Merleau-Ponty on 'body'. The article challenges body–mind dualism and the hyper-separation between nature and culture inherent in the Anthropocene. It calls for moving from the Anthropocene to Ecocene by making, what Küpers calls, an anthro-decentric transformation. The article ends by presenting implications for sustainable organizing where the role of the body and embodiment have a central stage. Moreover, the author notes that moving 'towards a more integral ecocene necessitates an ethico-political restructuring and transformation of contemporary organizations [...]', which '[...] involves analysing and questioning anthropocentric and interest-centric political practices and how they are used to accomplish and uphold power or control' (p. 12).

4. Conclusions

Based on the articles of this issue, it is evident to us that now is the time to move out from the Anthropocene. Just as the Anthropocene marked a global matter-energetic shift, this end of the human epoch also marks significant changes in the deep geological time of the Earth's history. Different temporal perspectives and rhythms will play a role in how the time after the Anthropocene will unfold. There is a need to begin to conceive time not only in anthropocentric terms, but more deeply and holistically, for instance, in terms of minerals, plants, and animals [2]. Thus, instead of merely seeking to save the world for future human generations, consideration and care of non-human objects, like rocks—constituents of the Earth—opens up a different time horizon, as the emerging geo-social literature cogently demonstrates [11,12].

A possibility is that the on-going mass movement of people and other earthbound beings will both be an outcome and reason for the new epoch. Furthermore, the travel of earthbound beings beyond the boundaries of Earth—the exploitation of space—is an issue calling for serious critical reflection. Finally, the mobility of deep geological formations of the Earth merits consideration as well; the movement of lithospheric plates has historically changed the course of life on the planet in a remarkable way. The trouble of moving, living, and dying together in the late Anthropocene necessarily brings about new practical and theoretical questions of power, as the recent formulations of 'geopower', for instance, highlight [13].

Finally, this scenario demands us to begin to develop post-anthropocentric ways of theorizing and doing research. There is a need to find a better balance in how all habitants of the earth are included in theory making, and involved in theoretical narration. A recent study of mosquitoes in the context of tourism provides a case in point [14]. Drawing from feminist new material literature, the authors suggest a post-anthropocentric approach that casts mosquitoes as fellow travelers, with which we are to live with—no matter whether we like it or not. The inevitable common worlding with multiple others, including tiny ones, needs to be better acknowledged in future theorizations.

The entanglement of humans with various creatures is strikingly visible now that the COVID-19 entangles with our bodies, bringing about major social, political, and economic consequences. It has forced us to consider many commonplace habits anew, including travelling. Ecologically, the pandemic might trigger a shift, even though the history suggests that 'business-as-usual' is plausible after the crisis [15]. Nevertheless, the pandemic efficiently highlights that the concept of the human mastery and control over other creatures is just an illusion.

Author Contributions: The authors' contribution to the special issue and this editorial corresponds with the order of the authors listed in this article. All authors contributed substantially to all tasks. All authors have read and agreed to the published version of the manuscript.

Funding: This research received no external funding.

Conflicts of Interest: The authors declare no conflict of interest.

References

1. Crutzen, P.J.; Stoermer, E.F. The Anthropocene. *Glob. Chang. Newsl.* **2000**, *41*, 17–18.
2. Heikkurinen, P.; Rinkinen, J.; Järvensivu, T.; Wilén, K.; Ruuska, T. Organising in the Anthropocene: An ontological outline for ecocentric theorising. *J. Clean. Prod.* **2016**, *113*, 705–714. [CrossRef]
3. Heikkurinen, P.; Ruuska, T.; Wilén, K.; Ulvila, M. The Anthropocene exit: Reconciling discursive tensions on the new geological epoch. *Ecol. Econ.* **2019**, *164*, 106369. [CrossRef]
4. Heikkurinen, P. (Ed.) *Sustainability and Peaceful Coexistence for the Anthropocene*; Routledge: Abingdon, UK, 2017.
5. Valtonen, A.; Rantala, O.; Farah, P. (Eds.) *Ethics and Politics of Space for the Anthropocene*; Edwar Elgar Publishing: Cheltenham, UK, 2020. (in press)
6. Ruuska, T.; Heikkurinen, P.; Wilén, K. Domination, Power, Supremacy: Confronting Anthropolitics with Ecological Realism. *Sustainability* **2020**, *12*, 2617. [CrossRef]
7. Mohorčich, J. Energy Intensity and Human Mobility after the Anthropocene. *Sustainability* **2020**, *12*, 2376. [CrossRef]
8. Rantala, O.; Salmela, T.; Valtonen, A.; Höckert, E. Envisioning Tourism and Proximity after the Anthropocene. *Sustainability* **2020**, *12*, 3948. [CrossRef]
9. Mazac, R.; Tuomisto, H.L. The Post-Anthropocene Diet: Navigating Future Diets for Sustainable Food Systems. *Sustainability* **2020**, *12*, 2355. [CrossRef]
10. Küpers, W. From the Anthropocene to an 'Ecocene'—Eco-Phenomenological Perspectives on Embodied, Anthro-Dece Ntric Transformations towards Enlivening Practices of Organising Sustainably. *Sustainability* **2020**, *12*, 3633. [CrossRef]
11. Clark, N.; Yusoff, K. Geosocial formations and the Anthropocene. *Theory Cult. Soc.* **2017**, *34*, 3–23. [CrossRef]
12. Yusoff, K. Anthropogenesis: Origins and endings in the Anthropocene. *Theory Cult. Soc.* **2016**, *33*, 3–28. [CrossRef]
13. Grosz, E. Geopower. *Environ. Plan. D Soc. Space* **2012**, *30*, 973–975.
14. Valtonen, A.; Salmela, T.; Rantala, O. Living With Mosquitoes. *Ann. Tour. Res.* **2020**. [CrossRef]
15. Ioannides, D.; Szilvia, G. The COVID-19 crisis as an opportunity for escaping the unsustainable global tourism path. *Tour. Geogr.* **2020**. [CrossRef]

Article

Domination, Power, Supremacy: Confronting Anthropolitics with Ecological Realism

Toni Ruuska [1,*], Pasi Heikkurinen [1] and Kristoffer Wilén [2]

1 Department of Economics and Management, University of Helsinki, 00500 Helsinki, Finland; pasi.heikkurinen@helsinki.fi
2 Department of Marketing, Hanken School of Economics, 00100 Helsinki, Finland; kristoffer.wilen@hanken.fi
* Correspondence: toni.ruuska@helsinki.fi

Received: 21 January 2020; Accepted: 24 March 2020; Published: 26 March 2020

Abstract: In this article, we study politics as domination. From our point of view, domination, especially in the Anthropocene, has had two vital components—power and supremacy. In order to dominate, one has to have power over others. In addition, the politics of domination, such as colonial oppression of Latin America, has required reasoning, justification, and legitimation, often connected to superiority (because of religion, society, or civilization) from the oppressor's end. Past and present political ideologies and programs, such as colonialism, imperialism, but also welfare state capitalism, neoliberalism and increasingly popular Green New Deal are examples of what we call "anthropolitics", an anthropocentric approach to politics based on domination, power, and supremacist exploitation. In contrast to the prevailing anthropolitics, this article discusses post-Anthropocene politics, characterized by localization and decentralization, as well as a steep reduction of matter–energy throughput by introducing a theoretical frame called ecological realism.

Keywords: anthropocentrism; Anthropocene; deep ecology; degrowth; domination; ecological realism; politics; post-Anthropocene; power; supremacy

---

1. Introduction

Throughout the history of civilization, politics has been a human-centered process. It has broadly meant the activities and practices connected to the governance of human affairs. Politics, in general, refers to a decision-making process about the most desired ends and means with local, global, and context-dependent variations, in which different interest groups interact in order to decide for things that they represent and/or are associated with. Since agrarian cultures [1], this process has ranged from very to quite exclusive, where the members of the ruling elites have made decisions for everyone else's behalf, and often at their expense. Slowly, over the course of centuries, the process of politics has claimed to become more inclusive, and now, in the 21st century, many humans are seen to have at least partial access or a possibility to participate in collective decision making [2]. Disregarding this assumed gradual increase in inter-human inclusiveness, politics is still very much about human relations. That is, humans decide for humans, without giving much thought and attention to other beings and ecosystems on Earth beyond narrow utilitarian aims.

During the past few decades, this narrow, exclusive, and anthropocentric conception of politics has been contested; for instance, it has been challenged in the fields of political ecology, environmental sociology, and political economy [3–5]. The worsening ecological crisis, which manifests in climate change [6,7], has also served as a wake-up call for some politicians and business leaders to realize that politics should not be exclusively about and for humans [8]. Especially in the light of the new geological epoch, the Anthropocene, it is clear that the realm of politics affects everything on Earth. Nevertheless, while there have been repeating calls to reform and revolutionize the human socio-economic organization—for instance growth-based capitalism [8–10]—human-centered political systems have prevailed and continue to gain support and widespread approval among the general public in the industrialized world.

This article claims that politics, and consequently human organization, has been (as well as largely continues to be) anthropocentric. This so-called "anthropolitics" is rooted in the domination of others, both humans and non-humans. Building on ecological realism [11–13], this article posits that the prevailing anthropocentric domination, power, and supremacy lie at the root of the ongoing ecological crisis. This article proposes that politics on all levels (from global to local) need to shift from so-called anthropolitics to post-Anthropocene politics, where the underlying drive will no longer be about domination but one of coexistence and inclusion.

Apart from introducing and developing ecological realism in relation to post-Anthropocene politics, another theoretical contribution of this article derives from the analysis of different varieties of anthropocentrism [14,15]. In addition, the article links them to the past, present, and future programs of anthropolitics (see the end of Sections 3 and 4). Anthropocentrism as a phenomenon is diverse, from which can be found ontological, epistemic, moral, and agential variants [15]. In addition to these, we want to deepen the understanding and complement the existing literature by adding spatial and temporal variants to scholarly analysis.

This article is structured as follows. In Section 2, we briefly introduce what we call anthropolitics and its proposed antidotal framework—ecological realism. Section 3 describes anthropocentrism, in the context of the Anthropocene, as a foundation for domination that entails power over other human beings and non-human beings and supremacist reasoning, justification, and legitimation that has had (as well as continues to have) catastrophic ecological and social consequences. Following this analysis and narrative, the article proceeds to describe how anthropocentrism has manifested in inter-human and non-human domination in the modern political realm. From anthropolitics, the article then turns to a closer discussion on ecological realism and post-Anthropocene politics.

2. Anthropocentrism and Ecological Realism

Aristotle famously remarked in *Politics* that human beings are political animals. Certainly, living together and coming together for joint decision-making is an apt characterization of one particular trait of the human species. This characterization surely is not exhaustive, but is it even an exclusive trait? An ant colony could be seen in this way as well, as could a pack of hyenas hunting and living together. But humans do differ, at least to some extent, from the rest of the animal kingdom in their ability for utopian thinking, negotiating different abstract future scenarios, and executing their collective plans by means of technology. Another distinctive feature in human organization is an ever more global and complex social realm that surrounds the phenomenon of politics. This realm, comprising of a multitude of social and political institutions and ideologies, has developed from organic societies (with limited division of labor and hierarchy) to industrial civilization marked by an extensive division of labor, socio-economic inequality and hierarchy [16]. Particularly in the context of city-states, a more careful way to characterize politics would be to argue that it is and has resembled a process of (class) struggle or a game of power. In this process, certain individuals, classes, and interest groups have sought to seize power and authority in order to rule and dominate other humans and non-humans for the benefit of this fraction often with supremacist reasoning, justification and legitimation. This diagnosis connects to the Nietzschean proposal of will to power, which is coupled with will to transform the biosphere [17] in order to extend domination (largely by means of increasing economic wealth).

It is evident that in the Anthropocene [18], the human species is the dominating animal, indicating that humans have the possibility and power to dominate others—an opportunity, which some humans use remorselessly and ruthlessly. In the age of human domination of the Earth, the scale of the human-induced natural destruction affects the living condition of almost all earthbound beings [6,19]. While these scientific findings are correct, it is important to add that it is a rather small portion of humans and their organizations, such as oil companies and other transnational corporations and particular nation states [20–23], which have been in the forefront of planning and implementing these destructive acts. From this critical observation, we can be deduce that the anthropogenic changes in the biosphere are also sociogenic [15]. In other words, they point to certain individuals (historically, and still today, mostly men), social order, and specific power relations and structures, as well as to particular economic organization [24–28]. This is the background of anthropolitics, where a powerful minority of humans dominate human and non-human beings for the assumed benefit of this minority.

To confront anthropolitics, a theoretical framework and philosophical lens is introduced in this article to discuss post-Anthropocene politics. This approach is called ecological realism. It draws its inspiration from various fields and traditions of thought, such as critical realism [29], historical materialism [7,30,31], deep ecology [32,33], and (eco-)phenomenology [34–36]. In ecological realism (see Section 4 for a more detailed portrayal), nature is not reduced to a resource, but sustainable life on Earth is rather based on ecocentric being, values, activities and practices. By adopting ecological realism, anthropocentrism is studied in relation to ontological questions (what exists, what is?), epistemological questions (what kind of knowledge we can have from nature, what kind of knowledge and science there is?), and axiological questions (what needs to be done to alleviate the ecological crisis, what is good and right, what is equality and what kind of actions lead to it?), as well as in connection to question of agency and space-time. In short, ecological realism could be considered as a framework for sustainable organization of human activities, i.e., a rough sketch to inform how humans should ideally act as particles of the greater whole, disregarding how we consider the nature of human beings and their behavior (rational, irrational, or something other).

According to ecological realism, humans are not superior to the rest of nature but are a species among others. Nature exists regardless of humans, and things happen in nature disregarding humans and their existence. Nature is not dependent on humans, although humans are dependent on nature [11]. Moreover, humans have their own distinct way of perceiving, acting, and being in the world, like other species do, and the consequences of human activities can be analyzed and estimated in relation to other species, non-living nature, and ecosystems (if only partially). In addition to these, human history has been, for instance from the point of view of economic and technological development, a cumulating and culture-dependent process, which is and has been utterly entangled and embedded within natural processes, which set the limits and frame for human activities and to their quality [7,31,37]. Accordingly, and in contrast to the current situation, human activities have to be in proper, sensitive and interactive relation to the non-human world, because all beings have the right to flourish and the right to species-specific life on Earth [11,33]. This does not, however, entail that ecocentric thinking and activities subscribe or lead to ecological determinism [38], reductivism [39] or mysticism [16], but the aim is rather to develop a holistic and processual way to conceptualize nature and frames for human activities and organizing [40].

Before we move on to anthropocentrism, we would like to note in agreement with Malm [7] that a single theoretical framework or its development does not necessarily solve ecological or societal problems; however, this does not mean that their development would not be worthwhile. In contrast, by offering a critique and alternative approaches to existing theories and concepts, we may be able to reach and discuss some of the problems that lie in contemporary academia and in current social theory and analyze their ecological and political ramifications. In other words, critique, theory, and theoretical work may contribute to fostering and formulating alternative practices and steer them toward ecological sustainability. The work on ecological realism is hopefully part of this process.

3. Anthropocentrism

In the political debates and discussions on the Anthropocene epoch [19,41,42], nature is seen primarily as a standing reserve or a resource pool to serve a wide variety of human technological and economic endeavors [9,17,43]. This kind of instrumentalism, a utility-based relation to nature, does not seem problematic from the dominant anthropocentric worldview [44] as nature is assumed to exist for humanity's sake [14]. However, this kind of human supremacist standpoint has been claimed to result in existential problems, such as climate change and the sixth mass extinction event [11,14,32,45–47].

An instrumental take on nature [9,43] has developed and intensified over the course of past centuries, which means that it originates from somewhere. Some events, historical developments, and culture-specific factors have influenced the human condition in such a way that, especially in Western cultures, human beings have been seen as separate and supremacist creatures from the rest of nature [45,48]. Even though it is difficult to differentiate these events and cultures specifically, that is, to argue from where exactly the instrumental perception of nature originates or "takes over," it is possible to locate it, at least to some extent, to humanist thinking and philosophy. Charles Taylor describes, in his book *A Secular Age* [48,49], a historical and cultural turn that he believes has led to the current human-centered and humanist culture. This turn took place in the beginning of the modern era, during the 17th and 18th centuries. According to Taylor, there was a shift in human thinking and self-understanding in relation to God and nature to what he calls Providential Deism: God's kingdom was not considered "ready" anymore, but unfinished and it was up to humans to make it whole again. In addition, there was a shift in focus and objectives, as the focus from divine and transcendent turned towards the mundane and ordinary: God's will was now considered to be in human flourishing. This turn, for Taylor [48] (p. 18) paved the way to secularism or exclusive humanism as he calls it, i.e., 'humanism accepting no final goals beyond human flourishing, nor any allegiance to anything else beyond this flourishing'. This entailed that humans began to perceive themselves increasingly as creators alongside God, or as continuators of God's creation, which arguably led to supremacist and separationist view of nature. Industrial and scientific revolutions then continued and amplified these developments by giving humans powerful resources and instruments to dominate and control each other and their extra-human nature [45].

Humanism, as a tradition, has been criticized extensively by some of the leading Western scholars [50–53]. Instead of discussing this critique, we want to emphasize the anthropocentric nature-relation in the humanist tradition. A common notion to both Lynn White Jr. [45] and Charles Taylor [48,49] is their attempt to explain why particularly in the Western civilization humans and nature have been seen as separate entities. In addition to this, they have sought to characterize nature-relation in Western cultures through instrumentality, domination and "economic value". In addition to these important contributions, anthropocentric perception of nature has been argued to be part of the Western philosophical tradition as well, for instance in Plato's *Timaios* dialogue, or in Descartes' body/mind dualism [40,54], and especially manifested in Bacon's instrumentalism [49] (pp. 230–232), [7] (pp. 209–210).

Famously, White Jr. [45] was among the first to link the root cause of the ecological crisis to anthropocentrism, which he saw deeply rooted in the Judeo-Christian theology and its underlying tradition of domination and supremacy. Devall and Sessions [44] took on their behalf White's ideas further as they identified anthropocentrism as the dominant worldview of industrialized societies. They [44] (pp. 65–66) wrote that "For thousands of years, Western culture has become increasingly obsessed with the idea of dominance; with dominance of humans over nonhuman Nature, masculine over feminine, wealthy and powerful over the poor, with the dominance of the West over non-Western cultures."

Krebber [55] (pp. 328–329) claims that, as productive forces developed, the alienation of humans from nature increased: 'The distance between culture and nature, mind and matter, subject and object, human and animal, grew until humans arrogated themselves the right to be above nature and finally broke the bond.' In his book *Eclipse of Reason*, Horkheimer [56] (p. 169) describes this development in history as reason infested by a disease pointing to instrumental reason seeking to manipulate and dominate nature. According to Horkheimer, instrumental reason has opted to adjust nature to meet human needs and aims, instead of seeking conformity with it, as it did in the past. Bacon [57] (p. 197) has famously crystallized this sea change in 17th century in following way: 'Let the human race only be given the chance to regain its God-given authority over nature, then indeed will right reason and true religion govern the way we exert it,' while later adding that nature should serve 'human affairs and interests' [57] (p. 329). Furthermore, Horkheimer and Adorno [58] (p. 1) wrote in *Dialectic of Enlightenment* that enlightened reason has 'aimed at liberating human beings from fear and installing them as masters.' This "reason" seemingly stems from the notion that nature is perceived as hostile to human life, which is then to be conquered, controlled and dominated in order, and paradoxically, to sustain humans [55] (p. 334).

Domination and modification of nature have extended vastly in their scope since the days of Bacon as even the weather system is now targeted for manipulation [55] (p. 331). Indeed, the instrumental perception of nature has been intensified by productivist economics and growth-based politics, as well as with the wide range utilization of fossil fuel infrastructure, which require extensive and expansive use of natural resources [23,40,59]. In Krebber's reading [55], Horkheimer and Adorno describe the modern era in *Dialectic of Enlightenment* as continuous struggle for the mastery of nature, in which thinking and reason function as primary means to achieve this goal. Compared to pre-capitalist and pre-modern domination, along with European Enlightenment and scientific revolution, a change in technique occurred that laid the ground for an expansion in massive scale in human's ability to control and manipulate nature. Whereas control of nature was previously approached through imitation (*mimesis*), nature was increasingly reduced to mere material reserve for satisfying human desires and needs [55]. Not even Marxism—probably the most prominent strand in humanist though in criticizing capitalism—has offered systematic critique, not to mention change of agenda, to anthropocentric domination of nature, as Marxist theory, research and politics have largely been in favor of industrialization and modernization committed to the ethos of progress and technological development [60–63].

It is important to remark, however, that not all humans are dominators equally. A more accurate way to put this would be to discuss of degrees of domination. Anthropocentrism should thus be seen to represent the human will or tendency to dominate [64] (p. 10), and its enactment: the domination. This tendency might be understood theoretically and psychologically as a trait that all humans may share—with different degrees in it—whenever humans see or find themselves as "better", "superior", "above", or "separate" from others (humans or non-humans) for whatever reason [14] (p. 309).

The enactment of domination has often been based on supremacist reasoning and justification. To be sure, other factors have influenced the relations of oppression as well, that is, slave trade might have been justified with supremacist arguments, but the primary reason for it was, arguably, monetary gain. Another point to consider is the fact that domination does not need to be justified in every case, especially when relations of oppression have been institutionalized (e.g., legislation concerning slavery, racial segregation, or gender discrimination). However, what domination always requires is power, that is, the will or tendency to dominate, as such, does not do anything if it is not coupled with power over something or someone. If humans do have the will or tendency, it is only, in fact, the power over something and/or someone that provides humans the ability and possibility to dominate. This power can be, for instance, political, technological, or muscular. In the Anthropocene humans dominate each other by means of wealth, military force, and with various political arrangements, and the non-human world especially by utilizing fossil-fuel-powered technology [15,23,27,28].

If we conceptualize anthropocentrism in this way, it is clear that we cannot dismiss inter-human concerns from it. This is because anthropocentrism is clearly a rationale, which some humans or interest groups adopt for their purposes [14] (p. 313) and thus, anthropocentrism 'does not put human beings at the center of the universe but only certain humans: those who, for one reason or another, choose to dominate others.' In more elaborate terms, de Jonge [14] (pp. 310–311) explains that

> Whatever class of social actors one identifies as being most responsible for social domination and ecological destruction (e.g., men, capitalists, whites, Westerners), one tends at the most fundamental level to find a common kind of legitimization for the alleged superiority of these classes over others and hence, for the assumed rightfulness of their domination of these others. Specifically, these classes of social agents have not sought to legitimate their position on the grounds that they are, for example, men, capitalists, white, or Western per se, but rather on the grounds that they have most exemplified whatever it is that has been taken to constitute the essence of humanness (e.g., being favored by God or possessing rationality).

By placing inter-human concerns at the spotlight of ecological problems does not mean that we are being anthropocentric. This should rather be considered as a focus, a fruitful starting point for understanding why humans dominate each other and non-humans, as well as to ask where the grounds for this are [14].

Now, if it is posited that anthropocentrism is a foundation for human domination, there also might be a need to extend the analysis to highlight different variants of anthropocentrism in order to deepen our understanding of the phenomenon. Heikkurinen et al. [15] review and distinguish ontological, epistemic moral, and agential anthropocentrism. In this article, two other variants are suggested to anthropocentrism, namely "spatial" and "temporal" (see Table 1). As noted by Heikkurinen and his co-authors, some social science–based Anthropocene discourses represent ontological anthropocentrism, which suggests that the existence of entities and beings on Earth are dependent on humans. In other words, existence and being are supposedly tied to language and human perception [7] thus indicating ontological exclusiveness for humans.

Table 1. Varieties of anthropocentrism.

Variant	Main Assumption	Reasoning Based on
Ontological	Existence is contingent on human existence	Ontological exclusiveness
Epistemic	Humans (or certain humans) are exclusive evaluators	Epistemic superiority and utilitarianism
Moral	Only humans (or certain humans) possess moral value	Moral superiority and exclusiveness
Agential	Humankind is an agent or only certain humans have (proper) agency	Agential exclusiveness, superiority and speciesism
Spatial	Earth is a space for humans	Exclusiveness, superiority and instrumentality
Temporal	Time is human time	Linear time, human rhythms, urgency and developmentalism

Epistemic anthropocentrism [15], in contrast, considers humans, or alternatively only certain humans, as the only source of value, or rather exclusive evaluators, thus accompanying them also with the "right" to make decisions from everyone else's behalf, namely because the others are not assumed to possess, for example, knowledge or understanding of what is best for them. From our point of view, this type of thinking often involves a following utilitarian reasoning: if a certain activity is thought to produce utility to a certain interest group, or to a society, nation, species, bioregion, technology development, etc., then this action is considered to be legitimate. In contrast to these two, moral anthropocentrism differs from them as humans, or alternatively only certain humans, are thought as the only ones possessing morality, or the only ones that possess inherent moral value among beings [15].

Agential anthropocentrism then has its focus on agency. On the one hand, it treats humankind as an agent, which can be considered problematic from an inter-human perspective, if not complemented with a culturally sensitive understanding on agency. Categorizing the whole human species into the same group overlooks historical context, class, power relations, and ecological impacts of individual humans, organizations, and different economies [15]. On the other hand, agential anthropocentrism may suggest that some humans are more equipped and superior agents than others (e.g., men over women, or civil over indigenous). Agential anthropocentrism may also falsely indicate that only human beings are capable of rational and/or cognitive behavior.

Spatio-temporal anthropocentrism refers to seeing time and space merely in human terms. Therefore, this approach is likely to lead to favoring human activities in terms of space and time. Spatially, the Earth and beyond are considered as space for humans. Humans currently occupy most of the habitable land on Earth for agriculture (50% of all habitable land), forestry, mining, infrastructure and dwelling [65–67]. Humans have also historically have made claims for ownership and passed laws to appropriate land exclusively to the use of humans [68–70], which of course neglects all other species. Also, the outer space is increasingly conquered and exploited by humans. Spatial anthropocentrism is, in the Anthropocene, one could say, complete and ubiquitous as one species dominates the use, ownership, and control of space, however, with much disproportion in inter-human organization and impact [71,72].

Besides the spatial element, temporal anthropocentrism is about considering time as human time. It guides and encourages specific actions with an anthrotemporal justification. Eco-modernist thinking and activities could be thought as an example of temporal anthropocentrism, that is, when humans seek solutions for example to alleviate climate change by relying on nuclear energy and other modern complex and resource-intensive technologies, these solutions are often legitimized by pledging to a sense of urgency to transform nature [17]. Another example of temporal anthropocentrism would be to perceive human nations and organizations following a certain type or path of linear temporal development to a supposedly higher stage of progress, for example from hunting and gathering to agriculture, to justify domination.

4. Politics of Domination

In critical political theory, e.g., Marxism [10] and world-systems analysis [2], politics is commonly understood as a process of struggle, where different classes, interest groups and those in power seek to fulfill their economic, social and ideological objectives. As Wallerstein [2] (p. 51) writes, there is a constant struggle over the allocation of the surplus value in the current capitalist world-system. In this class struggle, states have been central actors ever since the French Revolution, which brought two essential changes to the modern political order. The first one was political expression of the theory of progress, very much in line with the Enlightenment. The second was modern conception of sovereignty, which stripped absolute power from monarchs and transformed it into the legislature of the people [2]. Since then, politics have very much been about exclusion and inclusion: 'Those who were excluded sought to be included, and those who were already included were most often inclined to keep eligibility for citizens' rights defined narrowly, maintaining the exclusions' [2] (p. 52).

More generally, unequal power relations, for instance concerning indigenous livelihoods, the ownership of the means of production and allocation of surplus value, have produced recurring conflicts in which the people in power have attempted to dominate others [5]. Colonialism, imperialism, and neoliberal capitalism, for instance, could all be considered as manifestations of a dog-eat-dog world, in which the process and the outcome have been similar through ages—that is, conquest, domination, and exploitation of others. Also, the legitimation and justification of domination has remained quite the same, although it has had historical variations. Supremacist arguments regarding the conquest, domination, and exploitation of indigenous peoples have altered from "civilizing mission" to "progress" and "development," the exploitation of the non-humans has shifted from conquest and colonization to utility. In short, for humans, modern civilization has meant exploitation and homogenization in the name of civilization and progress, and for non-humans, it has mean instrumentality, commodification, and destruction due to alleged human superiority.

Below, we illustrate anthropolitics the modern era by highlighting on what grounds people, nations, and dominant classes have claimed their superiority and exploitation. This analysis is by no means comprehensive, but rather, it is an illustration of anthropocentrism, its reasoning, and its justification and legitimation.

The Spanish colonization of the Americas in the 16th and 17th centuries gave rise to a theological, political and ethical debate concerning the use of military force to acquire control over foreign lands [73]. This debate occurred in the framework of religious discourse that sought to legitimate the conquest with arguments of conversion and "salvation" of indigenous peoples. The British colonizers later used similar civilizing mission argument in the 19th century, for instance, in India. In the Americas, the Spanish conquistadores explicitly legitimated their actions in terms of religious missions, by bringing Christianity to the native peoples (cf. epistemic, moral, and agential anthropocentrism). Innocent IV, among others, argued that the use of force was legitimate if the indigenous peoples violated the so-called natural law, which led the Spanish to conclude that the habits of the natives, from nakedness and unwillingness to labor, were demonstrations of this inability [73].

In contrast to many influential philosophers, Diderot was one of the most vigorous critics of European colonization. For one, he challenged the common view that indigenous people benefitted from European civilization, and did not consider non-Westerners as primitive, nor more complex forms of social organization as superior. He also claimed that traders and explorers had no right to access fully inhabited lands—"the right to commerce" was commonly used as a justification by Spanish thinkers in the 16th and 17th century [73] (cf. spatial anthropocentrism). In contrast to Diderot, several influential philosophers, for instance in France and England in the 18th and 19th centuries took entirely opposite view and assimilated some version of the so-called developmental approach from the Scottish Enlightenment. According to it, all societies naturally transformed from hunting, to herding, to farming, to commerce; or from "savagery" to "barbarism" and to civilization, where the term civilization marked the highest point of (moral) progress [74] (cf. moral and temporal anthropocentrism).

The language of civilization, savagery and barbarism is common to many 18th and 19th century thinkers, including Karl Marx and John Stuart Mill. Mill thought that "savages" did not have the capacity for self-government (cf. epistemic and agential anthropocentrism). He thought that only commercial society could offer individuals a possibility to realize their potential for freedom and self-government, thus leading to a conclusion that colonizing states are acting in the interest of the less-developed peoples by governing them [73] (cf. moral and agential anthropocentrism). Marx also had similar ideas to civilizing mission thinking, so typical for Western Enlightenment and Eurocentric viewpoints, as Traverso [75] (p. 156) writes. Marx considered the colonial world as the periphery of the West, and opined that the West would determine its evolution, because of its own political orientation was "regressive" and "immature" [75] (p. 157). In articles written in the 1850s, Marx deals with the British imperial conquest in India and China, where he stigmatizes the violence of colonialism but does not put into question the legitimacy of British conquest in the name of a "superior civilization" (cf. epistemic, agential and temporal anthropocentrism). Marx writes that the British Empire acted as 'an unconscious tool of history' in India, and its mission was at the same time destructive and regenerative [75] (pp. 158–159). Similarly, Kohn and Reddy [73] note that Marx's thinking of British colonialist domination reflects the same ambivalence that he and Engels have in *The Communist Manifesto* toward capitalism: 'Marx recognizes the immense suffering brought about during the transition from feudal to bourgeois society while insisting that the transition is both necessary and ultimately progressive.'

From Tocqueville's thinking can be found more direct and exclusive political reasoning regarding domination, as he asserted that French colonies in Algeria would increase France's status in relation to its rivals. Unlike the proponents of the civilizing mission, Tocqueville noted that the brutal military occupation did not introduce good government or advanced civilization, but quite the opposite. Instead, he saw the French national interests as paramount and moral considerations subordinate to political goals [73]. Tocqueville's line of thought can also be considered to reflect the modern geopolitical power relations, where different nation states, economic organizations and other agents of capital are seeking to further their economic and political interests often at the expense of others [2,76].

From the late 19th century, anthropolitics and its underlying domination shifts from supremacist reasoning towards economic and political utility, as imperialism and colonialism change form due to capitalist and nationalist competition. In Imperialism: The Highest Stage of Capitalism [77], Lenin famously reinterpreted the term imperialism to refer to the age of concentrated capital and competition between nation states and their corporations. As Foster [78] explains, this new imperialist stage, beginning in the late 19th century was seen by Lenin as an outcome of growth of gigantic capitalist firms, which often had monopoly power, and the close connection they had to nations states from which they originated. This development resulted to the nationalist struggle to control and utilize the human and non-human beings and natural resources at the global scale, leading to inter-capitalist and nationalist competition and eventually to two world wars [78].

In general, Marxist scholars are quite unanimous on the notion that imperialism is inevitable result of capitalism [28,77,79]. This is because expansive capital accumulation is found to lead in constant search for new markets and opportunities for further accumulation, as already Marx and Engels foresaw in The Communist Manifesto [79]. As Moore [27] (p. 87) writes, capitalism repeatedly 'exhausts its sources of nourishment,' pertaining that capital accumulation depends upon finding new frontiers for appropriation. Hence, capitalist relations of appropriation eventually extend their grasp to novel territories by bringing industrialization and imperialism alongside with them [27] (p. 101).

In the 21st century, after many centuries of colonialism, imperialism, and capitalist appropriation in the name of progress, development, Christ, civilization, and affluence, the ecological impacts of humans' unequal affluence are indeed shocking. As a consequence of the "great acceleration" [80], every major ecosystem on Earth is in decline [9] (p. 46). A focal problem for ecological sustainability is that nature is perceived, in this era and in capitalism, not only as a limitless external entity, but also as barrier among others to be transcended, but above all as a resource to be exploited [27] (p. 95). Moreover, a reason why, for example, many of the citizens in affluent industrial societies are not able to see the full human dependence and embeddedness in nature, seems to be linked to their overall consumerist lifestyle, and capitalist economics and politics, which give the accumulation of wealth (or economic growth) the first priority in societal goal setting [9] (see ontological, epistemic and agential anthropocentrism). Indeed, as Krebber [55] (p. 333) aptly puts it: "the very idea of being in tune with nature in Western, capitalistic thought, is self-evidently equal to the subjugation of nature."

The subjugation of nature to human needs, wants and desires has been equal also for the later phases of capitalism in the 20th and 21st century. Although, the unprecedented economic growth in the post-war years resulted in a few somewhat successful examples from the human point of view, these manifestations of welfare capitalism, for instance the Nordic welfare states, were by no means free or excluded from global relations of oppression and capitalist competition, not to mention their ecological destructiveness. The Nordic welfare states, for example, were built around economic growth and expansive exploitation of natural resources, which meant that they reproduced the same instrumental outlook on non-human nature as their predecessors did, albeit with more equality and inclusion between humans [81] (cf. epistemic and spatial anthropocentrism). Yet, in front of the ecological crisis these societies have proven as incapable as every other industrialized society to shift their course from ecological devastation to sustainability [82]. Instead of changing the course to the needed degrowth policies and economics [83], the response is similar to other nations and transnational corporations: daydreaming about clean technology and ways to decouple economic growth from ecological impacts in the future. As Kolakowski [10] (p. 1180) fittingly noted, "The situation is one of a kind frequently met with in history, where people feel they have got into a blind alley: they long desperately for a miracle, they believe that a single magic key will open the door to paradise, they indulge in chiliastic and apocalyptic hopes."

For the last four decades, the politics of welfare capitalism have been overtaken by politics and economics of neoliberalism [76,84]. Neoliberalism has been both an ideology for the ruling elites and an instrument for transnational capital and its agents [59] (p. 93). On the one hand, neoliberalism has been an ideological and political campaign claiming that private sector is more effective than the public sector. Neoliberal policies have favored "free trade," financial de-regulation, as well as re-regulation, big banks and transnational corporations, and the so-called trickle-down economics. On the other hand, neoliberal politics and economics have targeted the internal problems of capitalism by seeking to rearrange the global political field of accumulation and economic markets to favor the movements of transnational capital, by securing capital flows, privatizing healthcare, infrastructure, education, and by capitalizing, proletarizing, and urbanizing Asia, Latin America, and Africa. In short, neoliberal politics and economics have been used as a vehicle to secure and reproduce the rule of transnational capital and its agents [76]. In this era, the non-human nature has been subjugated, perhaps more aggressively than ever before, to the iron rule of private economic wealth accumulation, as the non-human world has been valued only when it has served the interests of capital and technology development [27,28].

In addition to this, socio-economic inequality among humans has reached its new peak in the neoliberal era [71,72,81]. Forty years of neoliberal politics has ravaged the blue planet and meanwhile led to a political dead-end, in which a very small minority dominates others and collects the diminishing returns from the global capitalist market economy.

Before moving on, it is important to note again that the capitalist mode of production has to expand in order to reproduce itself [59,85]. This implies that this productivist organization is not only in antagonistic relation to nature [40], but also to all other modes of organization and cultures, including subsistence and self-sufficiency. Capitalist appropriation started from the enclosures and plundering of the commons [27,69] and has continued for the past centuries to exploit and dominate all life forms on an increasing scale, while piling up waste and ecological destruction. As Illich writes in *Shadow Work* [86] (p. 139), the modern era "can be understood as that of an unrelenting 500-year war waged to destroy the environmental conditions for subsistence and to replace them by commodities produced within the frame of the new nation-state." This war has been a circle of coercion and pauperization, where peoples, cultures and non-human beings are enforced to be part of the sphere of accumulation, as Snyder [87] remarks. He concludes by stating, "When the commons are closed and the villagers must buy energy, lumber, and medicine at the company store, they are pauperized" (p. 38). Where colonialism was originally justified with bibles and Christ, it continues with micro-credits, bottom of the pyramid economics and Western educational curriculum, to replace subsistence economies with capitalist growth and market economies.

The future projections of anthropolitics are manifested in initiatives such as the popular Green New Deal [88–91]. The Green New Deal (GND) gives non-empirically grounded promises of simultaneously circumventing the socio-economic inequality and ecological devastation with politics and economics of green growth [92,93]. This new deal does not entail the demolition of capitalism or productivism, but supposedly offers a more responsible, moderate and "green" version of it without questioning economic growth and the underlying politics of domination.

GND takes its inspiration from Roosevelt's New Deal from the 1930's, which aimed to lift the US economy from the Great Depression. In short and in contrast to New Deal, GND seeks to solve the climate crisis and socio-economic inequality with huge investments in renewable energy, energy efficiency, and clean transportation, while keeping wage levels and benefits intact [90] (p. 30). The GND is by definition a growth and reflating program, which is destined to produce more wealth, increase the standard of living, and to make the US the center of world's manufacturing again [88,89,91], although some prominent public figures have also called for curtailing consumption and the overall consumerist culture [90] (pp. 264–265).

Overall, the GND seems to rely extensively on the myth of decoupling, i.e., to the idea that the ecological impacts could be decoupled from economic growth [92,93]. In this sense, GND does not vary much from other modern green capitalist economic agendas, although the reduction of socio-economic inequality is a key component of it. However, questions of equality are only extended to the human realm. One indication of this is the sense of urgency, which is also one of the key selling points of the GND, as Klein [90] (p. 33) [91] amongst others notes: "we need every action possible to bring down emissions, and we need them now." This type of rhetoric is emblematic to temporal anthropocentrism—a type of thinking where time pressure is used to legitimate vast transformations in the biosphere [18]. Indeed, the most striking feature of GND politics is the mere absence of the non-human world. When Klein [90] (p. 53), for instance, declares that let us forge GND for everyone this time, she clearly means only humans, as if human well-being would somehow be independent from the rest of the nature.

5. Post-Anthropocene Politics and Ecological Realism

'As sure as impermanence, the nations of the world will eventually be more sensitively defined and the lineaments of the blue earth will begin to reshape the politics. The requirements of sustainable economies, ecologically sensitive agriculture, strong and vivid communal life, wild habitat—and the second law of thermodynamics—all lead this way.' [87] (p. 47).

In order for us to imagine an alternative mode of politics, which is capable of leading us out of the Anthropocene epoch, it is necessary to disengage from the anthropocentric ethos of Western thinking [15]. Instead of building political agendas on ungrounded optimism about human skills, technology development and dangerous productivist ideals limited to human purposes (as in the GND for instance), post-Anthropocene politics rooted in ecological sustainability must emerge. Focusing on inter-human relations clearly neglects the relevance of extra-human relations in both epistemic and axiological senses, but also from an agential perspective [15]. In contrast to this, and while a reality independent of humans is presumed [33,94,95], non-anthropocentric ethics, philosophy, and agendas [96] are essential in the search for ecological politics and economics rather than the prevailing techno-optimism and productivism.

In ecological realism—as a framework for ecologically sustainable post-Anthropocene politics and sustainable human organizing—nature is considered as a whole hosting all earthbound beings, bioregions and ecosystems. Nature does not reduce to any perception or discourse nor can be captured by any piece of writing since words and language only reach a part of the totality [33,34]. Conceptions of nature have historically often been place- and culture-dependent, even if it can be argued that all beings (such as grizzly bears, humans, basses, vipers, deer flies, silver birch, sun flowers, etc.) also share an experience of existence (whether it is being, doing, or something other). While these experiences vary in spatio-temporal context not limited to human space-time, the experience of nature and existence is arguably distinctive for humans in the Antarctic than it is from polyergus' point of view in the Amazon rainforest.

Humans have not had a particular or unitary conception of nature. Although it may be problematic to argue that a certain conception of nature has resulted in the ecological crisis, it can nevertheless be assumed that certain conceptions of nature (and consequently approaches to politics and economy) have supported or assisted the crisis more than others have, as well as prolonged its continuance. As an alternative to an anthropocentric approach, which we have argued to have led or assisted the events and activities that have produced the present ecological crisis, we present here a short summary of the key elements of ecological realism. Our interpretation of ecological realism [97] relies on critical realism [29], historical materialism [7,30,31,98], and deep ecology [11,32,33], but is also influenced by (eco-)phenomenology [34–36] and strong sustainability studies [99]. Based on these traditions and lines of thought, we have built a synthesis, which we apply in this article for outlining post-Anthropocene politics.

According to ecological realism, nature and its ecology exist in the absence of humans. Humans can obtain knowledge and information about nature, even if only partial and only subjective kind (to be sure, some understandings are considered more subjective than others). Nature's material, historical and biophysical processes set limits and frames for human activity. Within this frame and space of nature, humans have developed, in course of time, techniques and technology, and various political structures, as well as various modes of economic organizing, such as subsistence economies, feudal societies and fossil capitalism. As a general rule of thumb in ecological realism, human activities should be in an ethically inclusive and sensitive relation to rest of nature. By relying to these rather broad principles, we next present further tenets of ecological realism, a mixture of ontological, epistemological and axiological premises.

5.1. *Embeddedness, Equality and Dependency*

Nature and its ecosystems are finite entities [99–101]. Each part of each ecosystem is dependent on the whole [33]. Nature as a whole does not reduce to a part, which also means that nature, as a whole, is more important than its individual parts [11]. This includes humans who are embedded in the whole of nature. Humans are only a single species among other species—as important as the others, but not more or less important than the others. As human cultures receive a wide-variety of life sustaining things from ecosystems, such as air, water, soil, minerals, animals, plants and microbes [102], humans are matter-energetically dependent on the web of ecosystems, the whole of which nature consists of. More or less independent parts of the whole, such as a flock of birds or a human tribal community, are internal parts of a particular ecosystem, which makes them dependent on the whole for their survival. The ecosystem itself, however, is not dependent on its individual parts, which signifies a hierarchical relation between the part and the whole [11,15]. To be sure, human activities affect ecosystems, but the relation is asymmetrical—humans can have an effect on the climate by burning fossil fuels, which means that there is interaction between different entities, but this does not make the climate dependent on humans [7].

5.2. Historicity, Materialism, and Specificity

History is a matter-energetic process, as all earthbound phenomena are. Artefacts that are built today are likely to exist tomorrow, unless they are systematically destroyed. Even this will leave a thermodynamic mark in the closed Earth system. The universe also exists matter-energetically, while going through different phases, consisting of different and more or less tangible substances and compounds. Human history has co-evolved and co-developed in interaction with the non-human processes of the blue planet and its biosphere [37]. This has been a cumulative and matter-energetic process: existing generations have utilized the knowledge, skills and tools of former generations and developed or sustained them and then passed them on to the generations to come.

A focal characteristic of the human species is its capability and drive to manufacture tools for the search, cultivation and harvesting of food, habitation and clothing, as well as to make various artisanal objects to be part of social and cultural gatherings [7,98,103]. Although some other animals utilize simple tools in their activities, humans are the only ones that manufacture tools by combining different substances, and the only one who makes tools to make other tools [7] (pp. 165–166). In addition, the use of tools and the accumulating material flows are tied to a complex human-to-human interaction, as well as require high abstraction. When manufacturing and using a particular tool or natural resource, the shape of it, its form and purpose have been designed in advance. The more complex human organizations are, the more abstract the language games behind these designs. From these remarks, we can gather that, even though humans are a species among others and embedded and dependent on the surrounding ecosystems, humankind has its own specific ways of perceiving, being, and doing and a history with a species-specific impact on them.

5.3. Agency, Intrinsicality, and the Right for Existence

The desirability of the consequences of human actions can be estimated and analyzed with the help of ecological realism. As the Earth's climate is largely changing due to the fossil burning actions of certain individuals and organizations [7], destructive agency can thus be located to those responsible of the burning. Most of the carbon dioxide in the atmosphere is a result from the act of (both directly and indirectly) burning the fossil fuels. As it is now commonly accepted that the climate change is anthropogenic, this also means that we need theories, concepts and methods that seeks to analyze and separate actors and processes from each other [104,105].

In addition, human activities have to be in sustainable relation to the rest of nature—that is, human actions cannot occupy space in excessive manner to build roads, cities, and infrastructure or to cultivate and harness food because the excessiveness of the current human organizing pushes ecosystems and other beings towards collapse and extinction. Even disregarding this "call for balance" in human activities in space and time, other species do have the right to exist as they are, and for their inherent behavior and existence, as well as for flourishing [11,33]. In other words, all life forms have intrinsic value and right to exist [46]. For post-Anthropocene politics, this entails judgment and self-limitation [106], as humans do not have the right to reduce or destroy the diversity of other life forms with other reasons than fulfilling their primary needs and desires [107]. Based on this premise, the human organization should always be put in relation to other beings and ecosystems [11].

6. Discussion

To overcome the Anthropocene [15], but also domination of nature as a foundation of (industrial) civilization [16,108,109], reconciliation with the whole of nature is necessary [55] (p. 325). This, however, does not have to signify that nature and society are amalgamated, as Malm [7] notes in his critical review of post-humanist thought. While it is true that post-humanist scholarly work has been able to challenge the anthropocentrism that is prevalent in the humanist tradition, post-humanist thinking also contains elemental problems. As Malm [7] (chapter 2) notes in actor-network theory, or in hybridism, as he calls it, society and nature—or social and natural phenomenon—are mixed in each other in such a fashion that they have become "one". According to this point of view, things and wholes, such as a human being, a car, a concrete bridge and water flowing in a river are blended together into one quasi-object [104]. "Hybrid" is thus an actor-network or a collage, which is partly human-made or modified and partly of natural substance. What is striking from our point view is the fact that the effects and qualities of different parts and wholes are not systematically analyzed or separated [104,110]. This leads to a conclusion that there is no "society" and "nature"—only actor networks or hybrids.

According to hybridism, it is difficult to argue, which causes the climate to change, because hybridism obscures the cause-effect relations or withdraws from their analysis altogether [7]. This bears great relevance in the human-dominated geological epoch, where humans must learn to cope with their power and accept responsibility for their actions. It is vital that humans are able to estimate and analyze their actions and their ecological impacts (if only partially). It is of course true that matter-energetically nature and society are mixed and remain in constant interaction with one another [31]. The society is not external to nature, but embedded in it, which indeed, does entail a degree of hybridism. Yet, whereas hybridism is ontologically an interesting and welcomed standpoint, it is analytically problematic if the degrees of hybridity are not considered. Furthermore, hybridism also remains poorly fitted with politics and societal change if it does not distinguish different actors and actions, that is, differences are needed for making judgments and valuations [7]. Nature and society certainly are mixed, but to understand the ecological impacts of human organizations, to respect the diversity in nature, and to enable politics to change, a variety of different categories and units of analysis (including nature and society) are required [7,15,22].

In addition, ecological realism can be seen as an attempt to move beyond an either/or type of utopian thinking where humanism and post-humanism, or anthropocentrism and ecocentrism, are opposed [111]. However, this does not mean that we are sympathetic to anthropolitics or human domination in the Anthropocene, but rather, we argue that post-Anthropocene politics do not need to discard the whole humanist tradition or posthumanist hybridism on the condition that there occurs a significant shift toward more ecocentric values, practices, and structures (see Table 2 below).

Table 2. Comparison of humanism, post-humanism and ecological realism [14].

	Humanism	Posthumanism	Ecological Realism
Ontology	humans and nature	humans and nature mixed	humans as part of nature
Epistemology	subject-object dualism	actor-networks, hybrids and collages	parts and wholes
Axiology	human specificity and superiority	equality of actants	equality, specificity and inherency of beings

In any case, the call is to imagine, outline, and implement post-Anthropocene politics and frameworks that can guide the way out of the human-dominated epoch. From our point of view, this requires an understanding of nature and politics that is inclusive and preserves Earth's diversity. By developing ecological realism as a framework to guide post-Anthropocene politics, this article offers the following outline, which is based on the principles of ecological realism, and the critique of imperialism, developmentalism, productivism and the overall politics of domination:

1. **Localization and decentralization**: Ecologically sustainable post-Anthropocene politics is a local and de-centralized endeavor, in contrast to global anthropolitics. Whereas nation-states and other centralized political organizations have historically proven ill-suited to govern large areas of land, local and place-based (eco)cultures have been, due to their place-specific and inter-generational knowledge, better equipped to govern their activities in relation to their environment [16,112]. Centralized political structures naturally centralize power but are also distanced from local politics, knowledge, culture, context, and bioregion, which may lead to poor judgment and to favoring the surrounding areas of the power center.
2. **Reduction of matter–energy throughput**: For the sake of natural diversity and the continuation of diverse life on Earth, a key political issue to tackle is the reduction of matter-energy throughput as a remedy to the ecological overshoot of over-consuming nations and economies [72,83,113–115]. This can be done by means of policy and political steering toward negative economic and population growth. In practice, this could imply, for example, caps on consumption, large and unified conservation areas, maximum levels of income, carbon and resource throughput taxes, etc. Beyond these economic and political "fixes," the more general drive of post-Anthropocene politics is to dismantle industrial and growth-driven ways of life, indicating also the need to pull apart from capitalist and other productivist modes of organization [40].
3. **Conservation and restoration of non-human habitats**: Another starting point for post-Anthropocene politics would be to commit to the normative position that human activities should not destroy non-human habitats or over-crowd the planet's surface in their current intensity. Motives and rationales for this position can vary, but it seems obvious to the current anthropolitics encouraging economic and population growth is on the track to the sixth mass extinction [20,47]. To reverse this trend, the Earth needs time and space to heal. Recently, there has been discussion surrounding the 50/50 proposal [116], which is a promising starting point, but not a desired final outcome (as 50% percent reserved for one species does not seem just or meet the definition nor biophysical conditions of diversity). Ultimately, and according to deep ecology and ecological realism, humans have accept the notion that they are just one species among the rest and thus deconstruct their colonization of the Earth.
4. **Meeting primary needs and desires**: It is not anthropocentric to be motivated by human needs and desires, as long as they fit to the ecological whole, i.e., human actions do not compromise the continuance of diverse life. While humans need and desire diverse nutrition, shelter, and clothing to keep warm and social interaction with fellow humans and non-humans, humans currently have and do many things that are not related to these needs and desires. There are a lot of things that do not fall into this category, which could be considered as humankind's secondary and tertiary needs and desires (such as smartphones, airplanes, oversea vacations, being a celebrity), which in the age of global capital and modern technology seem to be quite commonly shared (by those who can afford, have access to, and have an opportunity to pursue them). From the perspective of primary needs, humans can claim the right to clear a field in order to survive and dwell. This way humans may occupy habitable areas from other living creatures, but the reasons can be considered to be more legitimate when they deal with primary rather than secondary and tertiary needs. It goes without saying that all life is thus requires some killing and dying [33] but what is important here are the reasons behind killing and dying, as well as the scale of these measures. When the human organization reaches a point where human activities are not connected to the primary needs and desires, this activity can be questioned, as it is likely to occupy space from other beings. In contrast to these, the needs and desires that are actually basic and primary show universal characteristics, but are also dependent on history, place, bioregion, and culture. Thus, it is proposed here that so-called primary needs and desires-based politics could be one option to provide human cultures a framework to curtail or draw limits to consumption, production and secondary/tertiary needs and desires, and consequently, lead to sustainable well-being [99,117–119].

5. **Reruralization and self-sufficiency:** Such primary needs and desires-based thinking will draw our attention to places and times where they can be fulfilled. Therefore, and in order to move beyond a system of generalized private ownership, the access to and availability of land are among key issues for post-Anthropocene politics to tackle and reformulate. Moreover, and even if urban environments may offer a multitude of opportunities for human interaction, it is a matter-energetic impossibility to produce enough food and energy to sustain life in cities. Hence, the proposed post-Anthropocene politics should call for reruralization, and the politics of self-sufficiency.
6. **Radical inclusiveness and equality**: Post-Anthropocene politics is about inclusion, non-discrimination, and equality. While anthropolitics has been historically characterized by patriarchal domination and political exclusion of women and people of color, as well as non-humans, ecologically sustainable and socio-economically just politics have to be able to include a multitude and wide-variety of voices and stakeholders in its processes. Thus, politics should no longer be about seeking private interests or about advancing particular societal arrangements and ideologies but about coexistence and radical inclusiveness, which seeks to balance the needs of various beings and to conserve and regenerate the conditions of diverse life.
7. **Detechnologization**: In a world of almost eight billion people, it is extremely difficult to organize human activities sustainably with modern and resource-intensive technology [120]. On the one hand, the resource-intensity of modern technology and the quantity of humans on Earth denotes to a return to lower degrees of technology, and on the other hand, it implies that in future more and more work will be manual labor. Therefore, limits and thresholds are set to human economic activities both in physical and ethical sense, which leads us to conclude that the post-Anthropocene politics is about combining meaningful and satisfying life while radically decreasing the matter–energy throughput of industrial societies and reconfiguring their ethical, epistemic and ontological underpinnings.

7. Conclusions

Anthropocene epoch is by definition a result of anthropocentric domination. Humans have become a dominating force on Earth by dominating fellow humans and non-human beings. In other words, if (especially certain) humans or human cultures would have considered each other equal, as well as their non-human companions, it is imaginable that Holocene would still continue.

Anthropocentrism, and its central manifestation, anthropolitics, is a historical and unequal phenomenon. In this article, we have identified six variants of anthropocentrism, which all contribute to human-to-human and human-to-non-human domination through supremacist reasoning, justification and legitimation, or with economic and political utilitarianism and instrumentality. Our understanding is also that each component of the presented anthropocentrism typology has had a part to play in advancing the anthropogenic changes in the biosphere that have led the Earth to the new geological epoch.

To be able to depart from the course of destruction, alternative organizing and mode of politics are needed. In this article, we introduced ecological realism as a framework for post-Anthropocene politics and sustainable organizing, which seeks to rearrange and reconcile human activities with their fellow humans and non-human beings, ecosystems and nature as a whole.

It is obvious that the post-Anthropocene politics, proposed from the ecological realism framework in this article, do not fit together with the contemporary anthropolitics, which are mainly geared toward human interests with instrumentalism, utility, and wealth accumulation. Indeed, an effective post-Anthropocene political program is an antithesis, and hopefully an antidote, to the modern consumerism and productivism. Since ecologically sustainable politics are not possible in the current techno-capitalist political and economic structures, it makes little sense to propose reformist policy suggestions regarding greening the neoliberal capitalism or politics of ("green") growth. At the moment, post-Anthropocene politics are utopian politics for many, which seem to exist only in the margins and outskirts of contemporary societies. Our aim in this article has been, in addition to our attempt to portray anthropolitics as the domination of humans and non-humans, to shed more light to this margin so that it could someday become the mainstream, and thus its practices would shine and flourish more widely.

Author Contributions: Conceptualization, T.R., P.H., K.W.; approach, T.R., P.H., K.W.; writing—original draft preparation, T.R., P.H., K.W.; writing—review and editing, T.R., P.H., K.W. All authors have read and agreed to the published version of the manuscript.

Funding: Toni Ruuska has received funding for the writing of this article from the Maj and Tor Nessling Foundation.

Acknowledgments: We wish to thank the editors and anonymous reviewers of the article for their time and effort—your remarks were invaluable and helped us developing the argument. We would also like to express gratitude to our colleagues in the Limbo Collective for inspiring discussions. The paper was scheduled to be presented in Sustainability Science Days 2020 in Helsinki, but the event was cancelled due to coronavirus (an aggressive non-human agent).

Conflicts of Interest: The authors declare no conflict of interest.

References

1. Gowdy, J.; Krall, L. The ultrasocial origin of the Anthropocene. *Ecol. Econ.* **2013**, *95*, 137–147. [CrossRef]
2. Wallerstein, I. *World-Systems Analysis: An Introduction*; Duke University Press: London, UK, 2004.
3. Catton, W.R.; Dunlap, R.E. Paradigms, Theories and the Primacy of the Hep-Nep Distinction. *Am. Sociol.* **1978**, *13*, 256–259.
4. Eckersley, R. *Environmentalism and Political Theory: Toward an Ecocentric Approach*; Suny Press: New York, NY, USA, 1992.
5. Martinez-Alier, J. *The Environmentalism of the Poor: A Study of Ecological Conflicts and Valuation*; Edward Elgar: Cheltenham, UK, 2002.
6. Steffen, W.; Richardson, K.; Rockström, J.; Cornell, S.E.; Fetzer, I.; Bennett, E.M.; Biggs, R.; Carpenter, S.R.; de Vries, W.; de Wit, C.A.; et al. Planetary boundaries: Guiding human development on a changing planet. *Science* **2015**, *347*, 1259855. [CrossRef] [PubMed]
7. Malm, A. *The Progress of This Storm: Nature and Society in a Warming World*; Verso: London, UK, 2018.
8. Klein, N. *This Changes Everything: Capitalism vs. the Climate*; Simon & Schuster: New York, NY, USA, 2014.
9. Foster, J.B. *Ecological Revolution—Making Peace with the Planet*; Monthly Review Press: New York, NY, USA, 2009.
10. Kolakowski, L. *Main Currents of Marxism: The founders, the golden Age, the Breakdown*; W. W. Norton & Company: London, UK, 2008.
11. Heikkurinen, P.; Rinkinen, J.; Jarvensivu, T.; Wilén, K.; Ruuska, T. Organising in the Anthropocene: An ontological outline for ecocentric theorising. *J. Clean. Prod.* **2016**, *113*, 705–714. [CrossRef]
12. Bonnedahl, K.J.; Caramujo, M.J. Beyond an absolving role for sustainable development: Assessing consumption as a basis for sustainable societies. *Sustain. Dev.* **2019**, *27*, 61–68. [CrossRef]
13. Ruuska, T.; Wilén, K.; Heikkurinen, P. Ihminen osana luontoa: Ekologinen realismi ja kestävä taloudellinen organisoituminen (Human as part of nature: Ecological realism and sustainable economic organizing). *Tiede Edist.* **2019**, *44*, 135–149.
14. De Jonge, E. An alternative to anthropocentrism: Deep ecology and the metaphysical turn. In *Anthropocentrism: Humans, Animals, Enviroments*; Boddice, R., Ed.; Brill: Boston, MA, USA, 2011.
15. Heikkurinen, P.; Ruuska, T.; Wilén, K.; Ulvila, M. The Anthropocene exit: Reconciling discursive tensions on the new geological epoch. *Ecol. Econ.* **2019**, *164*, 106369. [CrossRef]
16. Bookchin, M. *The Ecology of Freedom: The Emergence and Dissolution of Hierarchy*; AK Press: Chico, CA, USA, 2005.
17. Heikkurinen, P. Degrowth: A metamorphosis in being. *Environ. Plan. E Nat. Space* **2019**. [CrossRef]
18. Crutzen, P.J.; Stoermer, E.F. The Anthropocene. *Glob. Chang. Newsl.* **2000**, *41*, 17–18.
19. Barnosky, A.D.; Matzke, N.; Tomiya, S.; Wogan, G.O.; Swartz, B.; Quental, T.B.; Mersey, B. Has the Earth's sixth mass extinction already arrived? *Nature* **2011**, *471*, 51. [CrossRef]
20. Hornborg, A. *The Power of the Machine: Global Inequalities of Economy, Technology, and Environment*; AltaMira Press: Walnut Creek, CA, USA, 2001.
21. Hornborg, A. *Global Magic: Technologies of Appropriation from Ancient Rome to Wall Street*; Palgrave Macmillan: New York, NY, USA, 2016.

22. Malm, A. *Fossil Capital: The Rise of Steam Power and the Roots of Global Warming*; Verso: New York, NY, USA, 2016.
23. Bauer, A.M.; Ellis, E.C. The Anthropocene divide obscuring understanding of social-environmental change. *Curr. Anthropol.* **2018**, *59*, 209–227. [CrossRef]
24. Di Chiro, G. Environmental justice and the Anthropocene meme. In *The Oxford Handbook of Environmental Political Theory*; Meyer, J.M., Gabrielson, T., Eds.; Oxford University Press: Oxford, UK, 2016; pp. 362–384.
25. Lövbrand, E.; Beck, S.; Chilvers, J.; Forsyth, T.; Hedrén, J.; Hulme, M.; Lidskog, R.; Vasileiadou, E. Who speaks for the future of Earth? How critical social science can extend the conversation on the Anthropocene. *Glob. Environ. Chang.* **2015**, *32*, 211–218. [CrossRef]
26. Malm, A.; Hornborg, A. The geology of mankind? A critique of the Anthropocene narrative. *Anthr. Rev.* **2014**, *1*, 62–69. [CrossRef]
27. Moore, J.W. *Capitalism in the Web of Life. Ecology and the Accumulation of Capital*; Verso: New York, NY, USA, 2015.
28. Ruuska, T. Capitalism and the Absolute Contradiction in the Anthropocene. In *Sustainability and Peaceful Coexistence for the Anthropocene*; Heikkurinen, P., Ed.; Routledge: London, UK; New York, NY, USA, 2017.
29. Archer, M.; Bhaskar, R.; Collier, A.; Lawson, T.; Norrie, A. *Critical Realism: Essential Readings*; Routledge: Oxon, UK, 1998.
30. Marx, K.; Engels, F. *The German Ideology*; Prometheus Books: New York, NY, USA, 1998.
31. Foster, J.B. *Marx's Ecology: Materialism and Nature*; Monthly Review Press: New York, NY, USA, 2000.
32. Naess, A. The Shallow and the Deep, Long-Range Ecology Movements. *Interdiscip. J. Philos.* **1973**, *16*, 95–100. [CrossRef]
33. Naess, A. *Ecology, Community and Lifestyle*; Cambridge University Press: Cambridge, UK, 1989.
34. Merleau-Ponty, M. *Phenomenology of Perception*; Routledge: London, UK, 2013.
35. Toadvine, T. *Merleau-Ponty's Philosophy of Nature*; Northwestern University Press: Evanston, IL, USA, 2009.
36. Bannon, B.E. *From Mastery to Mystery: A Phenomenological Foundation for an Environmental Ethic*; Ohio University Press: Athens, Greece, 2014.
37. Soper, K. *What Is Nature? Culture, Politics and the Nonhuman*; Blackwell: Oxford, UK, 1995.
38. Bettini, G.; Karaliotas, L. Exploring the limits of peak oil: Naturalising the political, de-politicising energy. *Geogr. J.* **2013**, *179*, 331–341. [CrossRef]
39. Daly, H.E. *Beyond Growth*; Beacon Press: Boston, MA, USA, 1996.
40. Heikkurinen, P.; Ruuska, T.; Kuokkanen, A.; Russell, S. Leaving Productivism behind: Towards a Holistic and Processual Philosophy of Ecological Management. *Philos. Manag.* **2019**. [CrossRef]
41. Crutzen, P.J. Geology of mankind. *Nature* **2002**, *415*, 23. [CrossRef]
42. Waters, C.N.; Zalasiewicz, J.; Summerhayes, C.; Barnosky, A.D.; Poirier, C.; Gałuszka, A.; Cearreta, A.; Ellis, E.; Ellis, M.; Jeandel, C.; et al. The Anthropocene is functionally and stratigraphically distinct from the Holocene. *Science* **2016**, *351*, aad2622. [CrossRef]
43. Heidegger, M. *The Question Concerning Technology and Other Essays*; Garland Publishing: New York, NY, USA, 1977.
44. Devall, B.; Sessions, G. *Deep Ecology: Living as If Nature Mattered*; Gibbs Smith: Salt Lake City, UT, USA, 1985.
45. White, L., Jr. The historical roots of our ecologic crisis. *Science* **1967**, *155*, 1203–1207. [CrossRef]
46. Purser, R.E.; Park, C.; Montuori, A. Limits to anthropocentrism: Toward an ecocentric organization paradigm? *Acad. Manag. Rev.* **1995**, *20*, 1053–1089. [CrossRef]
47. Ceballos, G.; Ehrlich, P.R.; Barnosky, A.D.; García, A.; Pringle, R.M.; Palmer, T.M. Accelerated modern human–induced species losses: Entering the sixth mass extinction. *Sci. Adv.* **2015**, *1*, e1400253. [CrossRef] [PubMed]
48. Taylor, C. *A Secular Age*; The Belknap Press of Harvard University Press: Cambridge, MA, USA, 2007.
49. Taylor, C. *Sources of the Self*; Cambridge University Press: Cambridge, UK, 1992.
50. Heidegger, M. Letter on Humanism. In *Basic Writings: Nine Key Essays*; Routledge: London, UK, 1978.
51. Foucault, M. *The Order of Things: An Archaeology of the Human sciences*; Vintage Books: New York, NY, USA, 1970.
52. Foucault, M. What is Enlightenment? In *The Foucault Reader*; Rabinow, P., Ed.; Pantheon: New York, NY, USA, 1984; pp. 31–50.
53. Soper, K. *Humanism and Anti-Humanism*; Open Court Press: La Salle, MI, USA, 1986.
54. Bateson, G. *Mind and Nature: A Necessary Unity*; E. P. Dutton: New York, NY, USA, 1979.
55. Krebber, A. Anthropocentrism and reason in Dialectic of Enlightenment: Environmental crisis and animal subject. In *Anthropocentrism: Humans, Animals, Enviroments*; Boddice, R., Ed.; Brill: Boston, MA, USA, 2011.
56. Horkheimer, M. *Eclipse of Reason*; Oxford University Press: New York, NY, USA, 1947.

57. Bacon, F.; Rees, G. *The Instauratio Magna Part II: Novum Organum and Associated Texts*; Clarendon Press: Oxford, UK, 2004.
58. Horkheimer, M.; Adorno, T. *Dialectic of Enlightenment: Philosophical Fragments*; Stanford University Press: Stanford, CA, USA, 2002.
59. Ruuska, T. *Reproduction Revisited: Capitalism, Higher Education and Ecological Crisis*; Mayfly Books: North of England, UK, 2018.
60. Benton, T. Marxism and Natural Limits: An Ecological Critique and Reconstruction. *New Left Rev.* **1989**, *178*, 51–86.
61. Biro, A. *Denaturalizing Ecological Politics: Alienation from Nature from Rousseau to the Frankfurt School and Beyond*; University of Toronto Press: Toronto, ON, Canada, 2005.
62. Camatte, J. *The Wandering of Humanity, in This World We Must Leave and Other Essays*; Autonomedia: New York, NY, USA, 1995.
63. Wendling, A. *Karl Marx on Technology and Alienation*; Palgrave Macmillan: London, UK, 2009.
64. De Jonge, E. *Spinoza and Deep Ecology: Challenging Traditional Approaches to Environmentalism*; Ashgate: Aldershot, UK, 2004.
65. WWF. Living Planet Report: Aiming Higher. 2018. Available online: https://wwf.panda.org/knowledge_hub/all_publications/living_planet_report_2018/ (accessed on 25 March 2020).
66. IPCC. Climate Change and Land: An IPCC Special Report on Climate Change, Desertification, Land Degradation, Sustainable Land Management, Food Security, and Greenhouse Gas Fluxes in Terrestrial Ecosystems. 2019. Available online: https://www.ipcc.ch/srccl-report-download-page/ (accessed on 25 March 2020).
67. Ritchie, H.; Roser, M. Land Use. 2019. Available online: https://ourworldindata.org/land-use (accessed on 25 March 2020).
68. Polanyi, K. *The Great Transformation*; Beacon Press: Boston, MA, USA, 1968.
69. Perelman, M. *The Invention of Capitalism: Classical Political Economy and the Secret History of Primitive Accumulation*; Duke University Press: London, UK, 2000.
70. Wood, E.M. *The Origin of Capitalism. A Longer View*; Verso: London, UK, 2002.
71. Oxfam. An Economy for the 1%. It's Time to Build a Human Economy that Benefits Everyone, Not Just the Privileged Few. 2017. Available online: https://www.oxfam.org/sites/www.oxfam.org/files/fileattachments/bp-economy-for-99-percent-160117-en.pdf (accessed on 25 March 2020).
72. Ulvila, M.; Wilén, K. Engaging with the Plutocene: Moving towards degrowth and postcapitalist futures. In *Sustainability and Peaceful Coexistence for the Anthropocene*; Heikkurinen, P., Ed.; Routledge: London, UK, 2017.
73. Kohn, M.; Reddy, K. *The Stanford Encyclopedia of Philosophy*. 2017. Available online: https://plato.stanford.edu/archives/fall2017/entries/colonialism/ (accessed on 25 March 2020).
74. Kohn, M.; O'Neill, D. A Tale of Two Indias: Burke and Mill on Racism and Slavery in the West Indies. *Political Theory* **2006**, *34*, 192–228. [CrossRef]
75. Traverso, E. *Left-Wing Melancholia: Marxism, History, and Memory*; Columbia University Press: New York, NY, USA, 2016.
76. Robinson, W.I. *Global Capitalism and the Crisis of Humanity*; Cambridge University Press: Cambridge, UK, 2014.
77. Lenin, V.I. *Imperialism: The Highest Stage of Capitalism*; Penguin Books: London, UK, 2010.
78. Foster, J.B. *Late Imperialism: Fifty Years after Harry Magdoff's the Age of Imperialism*; Monthly Review: New York, NY, USA, 2019.
79. Marx, K.; Engels, F. *The Communist Manifesto*; Penguin: London, UK, 2002.
80. Steffen, W.; Broadgate, W.; Deutsch, L.; Gaffney, O.; Ludwig, C. The trajectory of the Anthropocene: The Great Acceleration. *Anthr. Rev.* **2015**, *2*, 81–98. [CrossRef]
81. Piketty, T. *Capital in the Twenty-First Century*; The Belknap Press of Harvard University Press: Cambridge, MA, USA, 2014.
82. Bailey, D. The Environmental Paradox of the Welfare State: The Dynamics of Sustainability. *New Political Econ.* **2015**, *20*, 793–811. [CrossRef]
83. Kallis, G. *Degrowth*; Agenda Publishing: Newcastle upon Tyne, UK, 2018.
84. Jones, D.S. *Masters of the Universe. Hayek, Friedman and the Birth of Neoliberal Politics*; Princeton University Press: Princeton, NJ, USA, 2013.

85. Magdoff, F.; Foster, J.B. *What Every Environmentalist Needs to Know about Capitalism*; Monthly Review Press: New York, NY, USA, 2011.
86. Llich, I. *Shadow Work*; Marion Boyars: Boston, MA, USA, 1981.
87. Snyder, G. *The Practice of the Wild*; Counterpoint: Berkeley, CA, USA, 1990.
88. Gunn-Wright, R.; Hockett, R. The Green New Deal. Mobilizing for a Just, Prosperous, and Sustainable Economy. New Consensus. 2019. Available online: https://s3.us-east-2.amazonaws.com/ncsite/new_conesnsus_gnd_14_pager.pdf (accessed on 25 March 2020).
89. Ocasio-Cortez, A. Resolution. Recognizing the Duty of the Federal Government to Create a Green New Deal. 2019. Available online: https://www.congress.gov/116/bills/hres109/BILLS-116hres109ih.pdf (accessed on 25 March 2020).
90. Klein, N. *On Fire: The (Burning) Case for a Green New Deal*; Simon & Schuster: New York, NY, USA, 2019.
91. Rifkin, J. *The Green New Deal: Why the Fossil Fuel Civilization Will Collapse by 2028, and the Bold Economic Plan to Save Life on Earth*; St. Martin's Press: New York, NY, USA, 2019.
92. Hickel, J.; Kallis, G. Is Green Growth Possible? *New Political Econ.* **2019**. [CrossRef]
93. Parrique, T.; Barth, J.; Briens, F.; Kerschner, C.; Kraus-Polk, A.; Kuokkanen, A.; Spangenberg, J.H. *Decoupling Debunked: Evidence and Arguments Against Green Growth as a Sole Strategy for Sustainability*; European Environmental Bureau: Brussels, Belgium, 2019.
94. Bhaskar, R.; Høyer, K.G.; Næss, P. *Ecophilosophy in a World of Crisis: Critical Realism and the Nordic Contributions*; Routledge: London, UK, 2011.
95. Vetlesen, A.J. *The Denial of Nature: Environmental Philosophy in the Era of Global Capitalism*; Routledge: London, UK, 2015.
96. Collier, A. *Being and Worth*; Routledge: London, UK, 1999.
97. Gorz, A. *Ecology as Politics*; Black Rose Books: New York, NY, USA, 1980.
98. Bhaskar, R. Materialism. In *A Dictionary of Marxist Thought*; Bottomore, T., Ed.; Blackwell: Oxford, UK, 1983.
99. Bonnedahl, K.; Heikkurinen, P. (Eds.) *Strongly Sustainable Societies: Organizing Human Activities on a Hot and Full Earth*; Routledge: New York, NY, USA; London, UK, 2018.
100. Georgescu-Roegen, N. Energy and economic myths. *South Econ. J.* **1975**, *41*, 347–381. [CrossRef]
101. Daly, H.E. *Steady-State Economics*; Island Press: Washington, DC, USA, 1991.
102. Starik, M.; Rands, G.P. Weaving an integrated web: Multilevel and multisystem perspectives of ecologically sustainable organizations. *Acad. Manag. Rev.* **1995**, *20*, 908–935. [CrossRef]
103. Mumford, L. *The Myth of the Machine: Technics and Human Development*; Harvest/HBJ Publishers: New York, NY, USA, 1967.
104. Latour, B. *We Have Never Been Modern*; Harvard University Press: Cambridge, MA, USA, 1993.
105. Latour, B. *Reassembling the Social: An Introduction to Actor-Network Theory*; Oxford University Press: Oxford, UK, 2005.
106. Kallis, G. *Limits*; Stanford University Press: Stanford, CA, USA, 2019.
107. Naess, A.; Sessions, G. A Deep Ecology Eight Point Platform. 1984. Available online: www.deepecology.org/platform.htm (accessed on 25 March 2020).
108. Perlman, F. *Against History, Against Leviathan*; Black & Red: Detroit, MI, USA, 1983.
109. Jensen, D. *Endgame, Vol. 1: The Problem of Civilization*; Seven Stories Press: New York, NY, USA, 2006.
110. Wapner, P. *Living through the End of Nature: The Future of American Environmentalism*; MIT Press: Cambridge, MA, USA, 2013.
111. Latouche, S. *Farewell to Growth*; Polity Press: Cambridge, UK, 2007.
112. Boehm, S.; Bharucha, Z.P.; Pretty, J. *Ecocultures: Blueprints for Sustainable Communities*; Routledge: New York, NY, USA, 1972.
113. Meadows, D.; Meadows, D.; Randers, J.; Behrens, W. *The Limits to Growth*; New American Library: New York, NY, USA, 1972.
114. Meadows, D.; Meadows, D.; Randers, J. *The Limits to Growth: The 30-Year Update*; Chelsea Green Publishing: Hartford, VT, USA, 2002.
115. Foster, J.B.; York, R.; Clark, B. *The Ecological Rift: Capitalism's War on Earth*; Monthly Review Press: New York, NY, USA, 2010.
116. Nature Needs Half. Available online: https://natureneedshalf.org/ (accessed on 25 March 2020).

117. Helne, T.; Hirvilammi, T. Wellbeing and Sustainability: A Relational Approach. *Sustain. Dev.* **2015**, *23*, 167–175. [CrossRef]
118. Büchs, M.; Koch, M. *Postgrowth and Wellbeing Challenges to Sustainable Welfare*; Palgrave Macmillan: London, UK, 2017.
119. Buch-Hansena, H.; Koch, M. Degrowth through income and wealth caps? *Ecol. Econ.* **2019**, *160*, 264–271. [CrossRef]
120. Heikkurinen, P. Degrowth by means of technology? A treatise for an ethos of releasement. *J. Clean. Prod.* **2018**, *197*, 1654–1665. [CrossRef]

 sustainability

Article

Energy Intensity and Human Mobility after the Anthropocene

J. Mohorčich

Lehman College, The City University of New York, New York, NY 10017, USA; joseph.mohorcich@lehman.cuny.edu

Received: 31 January 2020; Accepted: 12 March 2020; Published: 18 March 2020

Abstract: After the Anthropocene, human settlements will likely have less available energy to move people and things. This paper considers the feasibility of five modes of transportation under two energy-constrained scenarios. It analyzes the effects transportation mode choice is likely to have on the size of post-Anthropocene human settlements, as well as the role speed and energy play in such considerations. I find that cars, including battery-electric cars, are not feasible under a highly energy-constrained scenario, that buses, metros, and walking are feasible but will limit human settlement size, and that cycling is likely the only mode of transportation that would make suburbs possible in an energy-constrained post-Anthropocene scenario.

Keywords: energy; transportation; mobility; energy intensity; Anthropocene; cities

---

> Past a certain threshold of energy consumption, the transportation industry dictates the configuration of social space.
>
> —Ivan Illich, *Energy and Equity*

1. Introduction

People living after the epoch of human dominance will likely control a smaller fraction of the Earth's surface area, its material resources, and the other animals on and in it than they do today [1,2]. As a result, they will have less control over the energy flows and stocks on and in Earth, and thus less available energy with which to move and make things. One characteristic of the current organization of life on Earth is that human beings, in large part because of their exclusive access to underground reserves of fossil fuels and their control of the Earth's surface area, material, and other animals, direct a large amount of the energy flows on and inside the planet. The primary energy supply of all human societies is about 585 exajoules per year. This is small compared to the amount of energy in the form of solar radiation striking the Earth every day (about 7680 exajoules), but enormous compared to what is harnessed by other life forms and past humans. Human use of fossil fuel reserves is likely to decline in most post-Anthropocene scenarios. The exigencies of climate change, resource depletion, the difficulty and mounting inefficiency of extracting harder-to-reach reserves, breakdowns in the large-scale organization needed for energy extraction, or a combination of these and other factors offer plausible reasons to think so. Human control of the planet's surface and other animals is also likely to decline, whether because of shrinking human populations, breakdowns in organization required to organize and control large amounts of surface area, or—as documented later—declines in energy availability that make motorized transportation over long distances increasingly difficult. As a result of declining fossil fuel use as part of a broader trend of decreased control of the planet's material resources, the amount of primary energy produced by and therefore available to people is likely to decline. What would an energy-constrained world be like? What would humans still be able to do in this world and which capabilities would they lose?

Past human societies, which had less available energy than today's societies, provide some useful evidence about how to function in low-energy environments, and so it is possible to read accounts of earlier societies for guidance on this question [3,4]. Consulting past societies as models for future ones remains necessary but insufficient, however, for making considered observations about the possible paths of existing societies because (i) our understandings of past civilizations are imperfect, as the tendency of historians to revise these understandings, often fundamentally, every generation or so demonstrates, because (ii) post-Anthropocene humans will have some form of access to and knowledge about technologies and schemes for organizing society, food and goods production, government, and so on that pre-Anthropocene societies did not, and because (iii) it is unlikely that present societies would simply revert to past forms upon the imposition of past constraints, not least because societies change irreversibly as they move forward. History is not a time-reversible process. There is little evidence in the historical record of straightforward reversion as compared to, say, reversion with modifications [4]. Other work focused on the Anthropocene proper has discussed ways of conceptualizing an exit from the epoch and the material realities that occasion and condition such an exit [5,6].

One way of making observations about future paths is to start from any physical or material limits that create boundary conditions for what future societies will look like [7]. This paper represents one such attempt to work out these boundary conditions with respect to primary energy production dedicated to transportation, especially daily commuting. I estimate the energy, in gigajoules per person, that will be available for transportation under two energy-constrained scenarios. I then examine five modes of transportation (cars [electric and combustion], metros, buses [electric and combustion], bicycles, and walking) and calculate how much distance each can reasonably cover given certain energy constraints. I discuss the implications of these estimates for the mode choice, speed, and physical size of human settlements in a world after human dominance.

2. How Much Energy and How Much Distance?

Current primary energy consumption per person in US is about 326 gigajoules (GJ). Worldwide, it is closer to 82 GJ per person [8,9]. This number captures the electricity consumed from power plants (including imported and domestic coal, nuclear, natural gas, hydroelectric, geothermal, solar [thermal, concentrated, and photovoltaic], wind, wood, biomass), from burning petroleum products like gasoline and diesel, and from burning ethanol and biodiesel, as well as "non-combustion" fossil fuel use, as in construction, feedstock, and manufacturing [9]. Primary energy consumption, therefore, offers a reasonably good measure of the energy required to sustain the lifestyle and activities of a given population. Its increase or decrease, therefore, is a matter of general concern.

Any future decline in the energy available to humans could follow a gentle slope downward along which optimistic predictions about the ability of economies to transition to carbon-neutral energy sources like wind and solar come true quickly and thoroughly. In this world, humans would be able to settle at a steady state in which available energy is lower, but still high by historical standards. The decrease in energy available to human societies could also come quickly. Given that humans' most current and most lucrative energy-production techniques (e.g., longwall mining, deepwater drilling, hydraulic fracturing) tend to rely on human dominance and complexity, failures in the support systems these techniques rely upon could lead to rapid, non-linear decreases in available energy. To quantify and explore the differences between these two possible futures, I model two scenarios that I call demi-Anthropocene and post-Anthropocene. The first, demi-Anthropocene (demi-A), represents a partial or optimistic exit from the human-dominated geological epoch. It is characterized by the sort of transition to entirely carbon-neutral energy presented by Jacobson et al. [10]. Jacobson et al. suggest that a reduction in total energy demand of around 57.1% is consistent with a transition to entirely carbon-neutral energy systems. Estimates of reductions in energy use vary among different analyses [11] but are generally within ten percentage points of each other and of Jacobson et al.'s estimate. I use the round figure of 45% of available energy per person in the United States (that is, a 55% reduction) in the demi-A scenario as it captures the median of such work and avoids clothing itself in a false precision. The demi-Anthropocene scenario represents the sort of world brought about by rapid shifts to renewable energy alongside the widespread electrification of transportation and industry. This is the sort of situation that would involve nation-states and companies aggressively investing in renewable energy in the next three decades, closing coal and gas power plants, and phasing out internal combustion vehicles quickly and completely by 2050 or earlier. These changes would be rapid and thorough but would not radically change the constituents of the global economy or the total amount of energy available to human societies. Note, for instance, that because the demi-Anthropocene is based on 45% of the current US energy primary energy supply per capita, it would actually involve an *increase* in world energy supply per capita (from about 82 GJ to about 147 GJ per person per year). The demi-A scenario, therefore, can accommodate visions of increasing global standards of living for all but the most energy-intensive populations, who would have to reduce their energy demand by more than half, either through efficiency improvements or straightforward reductions in goods, services, and travel. This scenario would also involve some cooperation from non-human factors: to arrive safely at a demi-A plateau, worst-case scenarios like runaway feedback loops from widespread Arctic methane release will probably have to not occur, as they are likely to frustrate the supply chains, social systems, and technological improvements that will make providing 147 GJ of energy per person per year possible.

The second scenario, post-Anthropocene (post-A), is harder to model. This scenario represents a severe decline in available energy. The post-A considers a clear transition out of the human-dominated geological epoch and into a world in which human control of the planet's energy has largely slipped away. In this scenario, societies can no longer harness the energy available on the planet as efficiently or at the scales they do today and are consequently limited in their ability to organize and operate societies that demand high concentrations of energy. In climatological terms, such a scenario might coincide with, to take one example, the hotter Earth scenario presented by Steffen et al. [7], in which humans live on in a world much warmer (perhaps four to five degrees Celsius above preindustrial levels) and less hospitable to them. Unlike the demi-A scenario, this is a world in which global levels of available energy would shrink from 82 GJ to fewer than 17 GJ. Standards of living would fall for most people on the planet, either slowly or quickly. Travel, the consumption of goods, and the provision of critical infrastructure services would become more difficult and less robust. There are fewer ways of estimating the energy available in such a situation, because these situations are more distant and more different from our own. No matter how many studies of the Eocene or a "hothouse Earth" are conducted, knowing the quantity of available energy in such scenarios remains out of reach. Thankfully, this analysis does not rely on precise knowledge about the future paths of extreme warming or social collapse. This paper focuses on understanding how human transportation systems are constrained under low-energy scenarios, however such scenarios are arrived at. For these purposes, I set a benchmark value of 5% of current available energy for the post-Anthropocene scenario (that is, a 95% reduction in available energy from 2018 levels in the United States).

I mostly rely on US data in this analysis (noted when otherwise) because the United States represents the largest, most carbon-intensive economy whose figures on vehicle miles driven, total primary energy production, and other metrics tend to be comprehensive and accessible. I also start the analysis from US numbers because the US has often served as a vanguard for the energy-intensive patterns of behavior (personal car use, high amounts of meat consumption, home ownership, large-scale suburban housing) and the infrastructure systems that realize those actions (industrial farming, national highway systems, fossil fuel extraction and delivery, financial support for home ownership, a large military distributed across the planet) that make the most difference in the energy-use figures of a society on a per capita and absolute basis.

In the demi-A scenario, total primary energy per US resident would decline to 146.7 GJ, down 55% from 326 GJ today. In the post-A scenario, it would be 16.3 GJ, down 95% from today. Of course, not all this energy is spent on travel. From 2016 to 2018, transportation consumed 28.4% of total energy usage in the US [12]. (This is up from 23% in 1965 and 26.5% in 1990.) This 28.4% transportation share leaves 41.616 GJ to each person per year in a soft-transition demi-A scenario and a scant 4.62 GJ per person per year in a post-A scenario. The task, then, is to see how feasible each mode of transportation is under these energy constraints. A table (Table 1) summarizing the results of this work appears after the following discussion of cars, buses, metros, bicycles, and walking.

Table 1. Energy intensity, distance, and predicted effective commute range by mode.

	Energy Intensity (MJ per Passenger-km)	Distance/Year (km)	Distance/Year (km)	Effective Commute Limit (km)	Effective Commute Limit (km)
		Demi-A	Post-A	One-Way, Demi-A	One-Way, Post-A
Cars (ICE)	1.3	25,610	2896	35	3.97
Cars (BEV)	0.67	49,691	5516	68	7.6
Metros	0.1584	210,181	23,353	287.5	32
Buses (ICE)	0.54	77,067	6844	105.5	9.3
Buses (BEV)	0.29	114,803	12,744	157	17.5
Bicycles	0.105	317,074	35,200	434.3	48.2
Walking	0.218	152,719	16,954	209	23

3. Cars

The average American drives 21,688 km (13,476 miles) per year [13]. This represents an energy expenditure of 77.53 GJ [14]. Using a conversion factor of 34.34 MJ per liter of gasoline and an average fuel efficiency of 10.4 L per 100 km, as in Prokopiak (2012)'s work, but adjusting for unit conversion and distance. This figure, divided by the 92.58 GJ that is used per person for transportation in the US today, indicates that ~83.7% of the energy budget for transportation in the US goes toward car travel, which is broadly in line with US commuters' current modal share (76.6% of commuters drive themselves to work and another 9% carpool) [15,16]. The energy expenditure of a car that is driven 21,688 km per year outstrips the demi-Anthropocene energy budget by 86% and the post-Anthropocene energy budget by 1677%, or sixteen times over. Note that 6 GJ roughly represents the chemical energy of combusting 1 barrel (159 L) of crude oil.

Even if cars dropped to zero percent of the US modal share, the energy spent on all remaining forms of transportation (14.95 GJ) still far exceeds the travel energy budget in a post-A scenario (roughly tripling it). This implies that further reductions among modes of transit seen as low-carbon transit, like commuter rail and buses, will be necessary after the Anthropocene. It is possible that, under scenarios in which the calories available to human beings in the form of food are limited, strategies like reducing walking in favor of biking (biking is somewhere between two to four times as efficient as walking, and has around 30% lower carbon intensity per km, depending on diet) [17], reducing walking and biking together by, e.g., reducing the size of human settlements and increasing density, or some combination of these and other strategies will be necessary.

Under the demi-A scenario, distance driven would run up against a hard cap of about 32,012 km/year. This assumes a highly efficient internal combustion car, the most efficient of which can move one person one kilometer at a cost of about 1.3 MJ. Even this is likely impossible, as it requires directing every last joule of energy spent on travel into cars. A less unrealistic "soft cap" would fall around 25,610 km/year. (This number assumes the current balance of energy shifts modestly toward non-car travel, with car travel commanding a reduced, but still high, 80% of the travel energy budget.) This would limit most drivers to a bit over 70 km of total driving per day, effectively limiting the distance between workplace, school, and home to under 35 km (~22 miles), or from White Plains, New York, to the southern tip of Manhattan, or a little less than the driving distance from Machakos, Kenya, to downtown Nairobi. This would put downward pressure on American car use, but is unlikely to impose hard limits: the average round trip car commute is currently about 51 km. Much of Europe would come in under this limit, albeit not by much: drivers in the UK, for example, covered 34.8 km per day in 2013, consuming about half of what the demi-A scenario would allow [18]. (Chinese driving numbers are harder to come by, but data from Cox and Liu et al. indicate that kilometers driven in China remain ~42% lower than the US but are climbing rapidly [19,20].) However, these figures remain within the quite optimistic demi-A scenario. The numbers become formidable when considering the post-Anthropocene scenario.

In a post-Anthropocene scenario, the hard cap for driving would be about 2896 km per person per year. Again, this comes from redirecting 100% of the 4.62 GJ energy budget for travel toward car travel and is therefore almost certainly an overestimate. The soft cap for driving would be, using the optimistic 1.3 MJ per person per kilometer energy intensity figure, about 1035 km. As before, this assumes a reduction in the energy budget directed toward driving down to 80%. (In absolute terms, this is still quite high.) This scenario limits drivers to just 7935 m of driving per day, limiting the distance between home and commuter destinations to 3900 m or so. This is the distance from Manhattan's 1st Street to its southern tip, or the length of one large airport runway. These distances are more readily covered by foot, bicycle, scooter, or another form of low-energy intensity travel. This, obviously, leaves little room for car commuting in a post-Anthropocene scenario. Even electric cars, which are more efficient than internal combustion engine (ICE) cars, are too constrained under a low-energy scenario to provide travel beyond a few kilometers.

Battery-electric vehicles (BEVs) vary widely in their efficiency, as do internal combustion engine vehicles. A stringently efficient ICE car like the Volkswagen Polo expends about 1.3 MJ to move one kilometer. An efficient BEV, like the Nissan Leaf, expends about 0.67 MJ to move the same distance [21]. Using these efficiency numbers, if 100% of the light-duty fleet were converted to electric vehicles, the post-Anthropocene soft cap for one-way feasible commute distance would rise to a still-modest 7556 m per person per day. (Keep in mind that these numbers do not incorporate the energy costs or material demands of manufacturing cars.) A rough approximation of materials and energy expended can be gleaned by simply looking at the weights of the relevant vehicles: most light-duty BEVs weigh between 1600 and 2500 kg, ICE cars between 1100 and 2500 kg, bicycles 7–16 kg, and walking shoes 0.2–1 kg. The manufacturing energy required roughly scales with the total weight of each vehicle. Cars take around 73 GJ to manufacture per unit [22,23]. A bicycle, based on weight alone, should take about 0.47 GJ to manufacture. Note that, when spread over a ~12 year lifespan, the per-year amounts are relatively modest for both vehicles. Energy consumption from use is about ten times more important than energy consumption from manufacturing. Note, however, that if manufacture energy is included as part of the energy budget, one car, BEV or not, consumes five years of the post-Anthropocene energy budget via the energy demands of its manufacture alone, without being driven a meter. Note further that this energy-only analysis does not touch on the availability of rare earth metals and other materials needed to produce the amounts of BEVs needed for mass electrification, material in little demand by bicycles and walking shoes.) Hydrogen cars fare worse than BEVs on energy intensity metrics, falling somewhere between BEVs and ICE vehicles [24,25]. In anything approaching a post-Anthropocene world, cars will require commutes so short that other uses of transportation are better suited even if one considers factors other than energy intensity.

If cars have little use where they are not impossible, how do other forms of transportation, like metros, fare?

4. Mass Rapid Transit

I will examine mass rapid transit systems, or metros, next. Catling (1966) found that the London Underground expends 0.1584 MJ to move one passenger one kilometer, and (the physics of moving trains having changed little in the intervening years) contemporary sources do not disagree with him, generally finding between 0.09 and 0.35 MJ per passenger-km [26]. This is about one order of magnitude less than a very efficient ICE car and one-fifth of the energy required by a BEV. It falls about halfway between walking, which is less efficient, and cycling, which is more efficient. Under the demi-Anthropocene scenario of 41.616 GJ available per person per year for transport, distance traveled via metro would be subject to a soft cap of around 210,181 km/year. (This number, as with the soft cap calculation for cars, assumes that rapid transit consumes 80% of the travel energy budget.) This is a hugely generous travel allowance, allowing each user to circle the globe several times per year, or to cover 575 km per day. From an energy standpoint, rapid transit can accommodate arbitrarily long commutes. Travel distances would be limited by constraints other than energy, like time and geography. Note that these numbers remain within the optimistic demi-A scenario and become more constrained in a post-Anthropocene world.

The post-Anthropocene scenario allows for 23,353 km per year of travel using mass rapid transit, or 64 km of total riding per day, with 20% of the travel budget left over for other modes (which, in a system dominated by metro use, will be necessary to cover 'last-mile' scenarios and other gaps in the network. Note, too, that intercity rail transit is usually slightly more efficient than metros, allowing for moderate travel between cities under this scenario.) This travel limit is no longer arbitrarily high, and this has real-world consequences: commuters will not have the energy budget to commute by rail between Washington, D.C., and Baltimore (about 125 km roundtrip via MARC, which as intercity rail requires similar energy per km to metro systems) five days per week, for example, or from Potsdam to central Berlin (about 120 km roundtrip via S-Bahn), or between Shanghai and its outlying suburbs, or between Tokyo and its outlying suburbs.

Metros, unlike cars, consume sufficiently low levels of energy in a post-A scenario to be feasible for daily commuting within human settlements no wider than about 30 km, or between settlements no further apart than 30 km. Their consumption of energy in a demi-A scenario allows for arbitrarily long commutes (over 285 km each way).

5. Buses

I examine buses, both diesel and electric, next. Cushman-Roisin and Tanaka Cremonini find that standard diesel-powered urban buses use 0.4 MJ to move one passenger one kilometer if the bus carries 20 passengers, and 0.6 MJ if it carries ten [27]. New York City, Los Angeles, and San Francisco report an average passenger load per bus of about 17 [28]. Plugging an average passenger load of 17 into the above estimates yields an energy intensity of 0.54 MJ per passenger per km. This is very close to US Department of Transportation statistics (on which Cushman-Roisin and Tanaka partially rely), which, converted appropriately, yield 0.53 MJ per passenger per km [16]. This is about 20% more efficient than a BEV and 58% more efficient than an ICE car. Diesel buses carrying an average passenger load consume about two and a half times the energy consumed by walking and about five times that consumed by cycling the same distance.

Diesel buses would, under a demi-Anthropocene scenario with a travel energy budget of 41.616 GJ per person, be able to cover about 77,067 km per person per year, assuming, as before, that conventional buses would be consume 80% of the energy for travel. This is 211 km per person per day, allowing for a fairly generous maximum commute distance of 105 km or so. Under a post-Anthropocene scenario, the yearly limit would be 6844 km per person, or 18.75 km per person per day, for a maximum commute of about 9 km each way. This is a fairly restrictive scenario, falling somewhere between mass rapid transit and BEVs. Commuters would be unable to traverse the length of Manhattan or the width of Paris. Electric buses, however, are said to demand less energy per passenger-km than do diesel buses. Do they offer a workable transportation solution under energy constraint?

Electric buses employ a battery-electric drivetrain with regenerative braking. The efficiency advantages of battery-electric systems relative to combustion engines (like the diesel or clean diesel systems used by most buses) are compounded by the effectiveness of regenerative braking, which is able to recapture significant energy otherwise lost to brake heat in the start-and-stop use cases common to urban bus routes. Full electric buses remain more expensive than their diesel predecessors, but their adoption is widespread in China, where about 90% of new bus sales are electric, and major cities like Shenzhen only use electric buses. Uptake has been slower outside of China, though a variety of major cities have committed to electric-only bus fleets by dates like 2035 (San Francisco), 2037 (London), and 2040 (New York City).

Electric buses are indeed less energy-intensive than diesel buses. Electric urban transit buses consume between 0.18 and 0.4 MJ per passenger per km, depending on the analysis used [24,29]. I will use an average figure of 0.29 MJ per passenger per km. Under this assumption, distance traveled via electric bus would have a soft cap of around 114,803 km/year under a demi-A scenario. (This number, as with the soft cap calculation used for other modes, assumes that electric urban buses receive 80% of the travel energy budget.) In an electric bus system, commuters would be able to cover 314 km per day without exceeding the energy budget for transportation. While not limitless, this distance covers virtually all commute scenarios. Travel distance, for electric buses, would be constrained by factors other than energy.

A post-Anthropocene scenario allows for 12,744 km/year of travel using battery electric buses, or 35 km of travel per day, with 20% of the travel budget remaining over for other modes (which, as before, will be necessary to cover 'last-mile' scenarios and other gaps). This travel limit allows for most travel within cities, few of which are wider than 17.5 km, but remains somewhat restrictive for most other scenarios. Travel between Flushing, Queens, and lower Manhattan five days per week would exceed the energy available for transportation in such a model. Nor could anyone commute between, say, Cairo and New Cairo under these constraints. In fact, commuters would be hard pressed to commute from the suburbs of all but the most compact major cities into their central business districts.

6. Bicycles

Other than some electric kick scooters and solar-powered vehicles built for specific contests, like Solar Impulse 2 or the Nuna series, the bicycle is the most energy efficient means of transportation known. This includes the animal kingdom, human-developed means of transportation, and every form of transportation under consideration in this study. (Bicycles also demand less manufacture energy than every form of transportation considered in this study aside from walking.) A person on a bicycle, moving at 16 km per hour, consumes between 0.095 and 0.115 MJ per kilometer per person, depending on the weight of the rider, bicycle, and the rider's position (upright positions are less aerodynamic than lower positions and therefore require more energy at the same speed) [30,31]. I will use a figure of 0.105 MJ, because this is the best estimate for a human of global average adult weight (about 62.3 kg).

The soft cap for bicycle travel (using the same 80% allocation as before) under a demi-A scenario is 317,074 km. This is high. Currently, the world record for most kilometers traveled in a year on a bicycle is 122,432, which was accomplished by biking twelve or more hours per day for a year. A bicycle commuter would have the energy budget to travel 867 km per day. This limit is mathematically impossible to exhaust: 16 km per hour for 24 h is only 384 km. Even much faster cyclists, consuming energy and distance at a faster rate, would be hard pressed to exceed this limit: the world record for cycling for 24 h straight was set at 896 km in near-perfect conditions at an airport in Berlin. So, the distance limits for bicycle commuting are unconstrained by available energy under a demi-Anthropocene scenario. A great deal of the energy saved by bicycle commuting could eventually be reallocated to other uses, like manufacturing, medicine, or agriculture.

The soft cap for bicycle travel under a post-Anthropocene scenario, allocating 80% of energy to cycling, would be 35,200 km per person per year. This allows for 96.4 km of travel per person per day, or an effective commute distance of 48 km on the high end. This is the most generous allowance of any post-Anthropocene mode. Commutes between Texcoco and Mexico City, or Plano and Dallas, would remain possible seven days per week. Commutes between downtown Dallas and downtown Fort-Worth would remain possible five days per week, as would travel between famously sprawl-friendly Houston and virtually all of its outlying suburbs. The few examples that would be energy limited are already limited by simple practicality. Cycling from Irvine to downtown Los Angeles passes beyond the edge of the energy frontier but takes between four and five hours anyway—hardly a practical commute.

In all but the most extreme cases, the bicycle keeps the suburbs within reach of the central business districts to which they are nominally attached. Ironically, given the historical intertwining of the car and the suburbs, the bicycle is the only mode of transportation yet studied that would make something like true suburbs possible in a post-Anthropocene world.

7. Walking

Walking, because it does not take advantage of a wheel and ball bearing system to conserve forward momentum, is less efficient than riding a bicycle. Because it does not require any external weight or complication, it is more efficient than almost all forms of motorized transportation (electric kick scooters are, under the right circumstances, a possible exception). Combining the global average adult weight (62.3 kg) with evidence from McArdle (2000) on the energy demanded by walking, running, and cycling yields an energy consumption rate of 0.218 MJ per pedestrian per km [31].

A demi-Anthropocene society in which 80% of the transportation energy budget is allocated to walking can supply the energy for each of its members to cover about 152,719 km on foot per person per year. This is 418 km per day, or more than most humans can walk in several days. As expected, energy is not a significant constraint for walking under a demi-A scenario. Practical limitations remain more important than considerations of energy expenditure.

Under a post-A scenario, members of the same walking-dominant society would be able to cover 16,954 km per year, or about 46 km per day. This at the limit of what an experienced walker can cover in one hard day of hiking. Walking is unlikely to be constrained by energy after the Anthropocene. Instead, it will be constrained by the usual things that hamper walking long distances in any situation: walking is slow (about 3–7 km/h), and so it takes a long time to get anywhere further than a few thousand meters away, you are exposed to the elements while you do it, it is tiring, sensitive to terrain, and a burdensome means of transporting cargo.

8. Motorized Versus Non-Motorized Transportation

Note that cars, metros, and buses draw from energy stores external to themselves, either via electricity or the pumping of liquid fuels like diesel and gasoline. Human-powered transportation, like biking and walking, draws on the metabolic processes of human beings and, therefore, from the food humans eat and stored energy in the form of fat. The unit we have been using so far, the megajoule, is equal to about 239 kilocalories. A cyclist who uses 0.1 MJ to travel one kilometer will require about 24 kilocalories of food, or a bit less than one walnut, to do so. This means that, unlike the motorized forms of transit, the diet of the person walking or cycling matters for the energy efficiency and environmental effects of these forms of transit. A cyclist who mostly consumed beans, wheat, and corn, for example, could have an environmental impact as low as one-twentieth that of a beef-fed cyclist [32,33].

Please note that the transportation budgets used in this paper are derived from primary energy production in the United States. These numbers carry more or less straight over to motor vehicles, which use primary energy sources like gasoline, diesel, and electricity. However, primary energy production does not directly include kilocalories consumed per person per day or some equivalent measure. The overall numbers for food energy are not too hard to calculate (e.g., US residents eat about 3300 kilocalories per person per day, which is about 5 GJ of food energy per person per year), but for consistency, this paper assumes that for societies in which large-scale conversions to non-motorized forms of transportation occur, the pool of primary energy currently used for transportation will, in some form, still be drawn on by non-motorized forms of transportation. This may involve, as a practical matter, transferring material and power from centralized electricity production or fossil fuel extraction toward increased kilocalorie production, but the underlying energy budget will remain similar.

There are, finally, several reasons for preferring mechanically simpler and physically lighter forms of transportation in a post-Anthropocene scenario aside from energy intensity. Cars, buses, and trains, whether combustion or battery-electric, require extensive infrastructure in the form of electrical grids, power plants, substations, mining, machining, and so forth. Bicycles and shoes require no fueling infrastructure beyond potential increases in kilocalorie supply (i.e. growing and transporting slightly more food). They also require lighter and fewer manufacturing, supply chain, and raw material components for their production and maintenance. These economies can be Fermi-estimated by looking at the curb weight of bicycles and shoes against the curb weight of motorized vehicles. Bicycles weigh between two or three orders of magnitude less than cars, buses, and metros do (even when divided by the average passenger load of each). Shoes weigh about four orders of magnitude less than those vehicles. We should expect, at a rough cut, a bicycle to demand about 1/500th the materials and complexity that a car, bus, or train would (on a per passenger basis). Similarly, we should expect to be able to produce thousands of pairs of shoes with the materials and manufacturing required for one car. In a post-Anthropocene scenario in which materials and supply chains are limited or damaged, these material economies are likely to matter a great deal.

Considerations of equity also offer reasons to prefer simpler, lighter means of transportation. Because cars, buses, and trains are more expensive to produce and use than bicycles and shoes (by orders of magnitude similar to their weight differences), they are more expensive for people to use. This is likely to translate preexisting inequalities into the transportation system, as is currently common. Fewer people can afford a car than can afford a bicycle. Whether the costs for complicated, resource-intensive transportation are paid upfront, financed via debt, or amortized via ticket sales, road taxes, and insurance, they fall heaviest on those least able to pay. The only exception to this fact would be a society that used some form of progressive taxation to subsidize motorized travel (either via public transit or, say, by subsidizing private car ownership). (This same tax-and-subsidy scheme, however, could be used to make cycling or walking more affordable, likely to greater effect.)

9. Implications and Findings

Changing a mode of travel's speed modifies the relationship this form of travel establishes between energy consumption and distance. At higher levels of speed, more energy is expended resisting the movement of air. As a result, running the same modes of transportation more slowly is one way of eking more distance out of a limited energy budget. Hastening a mode of transit is, conversely, a way of trading energy for time. A cyclist traveling at 15 km/h requires 30 watts (or 30 joules per second), less than half of which is spent on air resistance [30]. A bicycle traveling at 30 km/h demands 148 watts. More than two-thirds of this power goes to moving air out of the way. (This assumes that the rider and bicycle have a combined weight of 75 kg and an average-sized frontal area.)

Pedestrians experience similar costs associated with speed, as do motor vehicles. A rough application of the drag equation indicates that an intercity bus moving at 30 km/h demands 1875 watts (joules per second) to push through the air (assuming a fairly typical drag coefficient of 0.6). ($F_d = \frac{1}{2}pu^2C_dA$, where p is the density of air, u is flow velocity, C_d is the drag coefficient of the vehicle, and A is its frontal area. Drag force (F_d) can be used to calculate the power needed to overcome drag at a given velocity using $P_d = \mathrm{F}_d{\cdot}\mathrm{v} = \frac{1}{2}pv^3AC_d$, where v is velocity, p is again air density, v is velocity relative to Mach 1, A frontal area, and C_d drag coefficient. Note that the power needed to resist drag increases with the cube of velocity.) The same bus moving at 60 km/h loses 15,000 watts to air resistance, or 15 KJ per second. Decreases in energy efficiency via higher speed reduce the effective range for all mode types under constrained-energy scenarios, with the exception of walking, which is still mostly limited by factors like time and fatigue. If air resistance accounts for, say, 40% of the energy expended by an intercity bus running its route, and the energy spent overcoming air resistance increases eightfold when speed doubles from 30 km/h, something like a 35% reduction in range can be expected from running buses closer to 60 km/h as opposed to 30 km/h. Combining the range numbers from the post-A scenario with the increased air resistance estimates means that diesel buses would have a new effective commuting range below six kilometers each way and electric buses would be unable to cover commuting distances more than 11 or 12 km each way. Because the energy required to overcome air resistance increases non-linearly, further speed increases shrink this range to a pinprick. In an energy-constrained system, speed is ruinously dear.

Illich (1974) identifies the severe demands that speed makes on energy as a structuring feature of societies, especially those that provide means of transportation that go faster than about 24 km/h [34]. Illich points out that "[i]n traffic, energy used over a specific period of time (power) translates into speed," because of which certain levels of speed will demand sufficiently high concentrations of energy and physical separation from other people and machinery as to change the physical and social organization of a human settlement. He argues that "the critical quantum" of speed is about 15 mph, or a bit over 24 km/h, and that "[w]herever this limit has been passed [a] basic pattern of social degradation by high energy quanta" can be observed [34]. After this limit was breached by public or private vehicles, "equity declined and the scarcity of both time and space increased. Motorized transportation monopolized traffic and blocked self-powered transit" [34]. As we have seen, something important inheres in the relationship between speed and energy intensity, but a more detailed exploration of these claims is needed. This paper has attempted to analyze the specific energy intensity of different modes of transportation in a way that prices in their average velocities but also takes into account their various efficiencies and energy demands as really-existing systems of movement. This information allows for a reevaluation of Illich's observation. We do see some correlation between the modes of transportation with high energy intensities and high speed (Table 2).

Table 2. Energy intensity and typical effective speed.

	Energy Intensity (MJ per Passenger km)	**Speed Range Usually Achieved in Cities**
Cars	0.67–5	30–95 km/h [35]
Metros	0.1584	25–85 km/h
Buses	0.29–0.54	13–85 km/h
Cycling	0.105	13–32 km/h
Walking	0.218	3–7 km/h

This correlation is somewhat jagged, however, with buses fairly inefficient for speed traveled, and metros and bicycles fairly efficient. Cars remain much less efficient (even for their speed) than other ways of getting around. Illich's dictum captures an important insight about the relationship between speed, energy, and the physical and social organization of cities, but probably remains too simple. A cyclist or metro train at 30 km/h commands a different amount of power, and therefore a different organization of energy, than does a car traveling at that same speed. Both are over Illich's speed limit. But under the energy-constrained scenarios I have considered, a car cannot be sustained at that speed for the length of an average commute, while a train or bicycle can. Moreover, the amount of infrastructure and space that must be dedicated to cars traveling at 30 km/h through a city is much higher, on a per-passenger basis, than what must be dedicated to make it possible for subway cars and bicycles to travel 30 km/h through the same city. The reorganization of space to allow for increased energy-intensity travel has certain social effects that urbanists, historians, environmental scientists, and others have documented: neighborhoods delimbed by controlled-access highways, city centers sheared by surface parking, pedestrians pushed back by road widening, playgrounds dusted with $PM_{2.5}$, those without cars separated from job centers and schools by roads, those with cars tethered to the fueling and maintenance of cars, and so forth. All of this is to say nothing of atmospheric carbon at 412 ppm and rising, ocean pH >8.1 and falling, and September Arctic ice extent <4 million km^2 and shrinking [36–39].

Under future conditions of energy constraint, Illich's posited relationship between energy intensity and the organization of cities may begin to work in reverse. Higher levels of speed will not drive increased extraction and the reordering of social and physical space. Instead, decreased extraction will put downward pressure on speed and will in its own way change social space. As concrete examples, I have listed certain commuting routes that will be difficult or impossible for simple energy-expenditure reasons in a post-Anthropocene scenario. No more Flushing to Manhattan, no more Machakos to Nairobi, and so on. One observation remains common to every scenario and mode of transportation under discussion: because humans will be restricted to distances that are smaller and will have to cover these distances more slowly, human communities will experience pressure to become geographically smaller. This means that these communities will have fewer people, those people will be more densely packed, or some combination of the two. If humans are reluctant to give up the economic, social, and cultural benefits that come with living in organized settlements with many other people [40], they will have to choose density, probably with effective commuting distances shorter than 17 km or so if using electric buses, 30 km using rapid transit, and 48 km using bicycles. The shape and extent of communities may conform to something of a weighted average of these figures, dependent on how much of a role the various modes of transportation play and how much energy is allotted to each.

(There is another set of scenarios, of course, in which humans do give up on the agglomerative effects of living together and live in relatively isolated semi-rural communities with no complex system of transportation to speak of because no one has any commuting to do beyond walking in the immediate vicinity of one's own settlement. I do not model this future, although I can say from the work here that, while bicycles are likely to be effective in these futures, it is hard to produce bicycles without some degree of agglomeration, trade, and design.)

The analysis in this paper suggests one motivating factor for policy change and about six concrete normative implications. The motivating factor comes from the fact that a demi-A scenario is probably not that bad and a post-A scenario is quite bad. A lot of useful transportation technology (to say nothing of other socially useful goods not considered here) will be useless or severely limited in a world where the total energy available per human being is under 17 gigajoules. Human societies should make every effort to land on a higher energy plateau, around demi-A, than on a lower one. This objective remains possible, but it will require aggressive efforts to lower emissions and some degree of luck. The more rapid and effective the effort, the less luck will be needed. The first normative implication is that societies ought to plan for lower-energy states in ways that are far-reaching and detailed. Transportation is an obvious area, and one severely affected by energy constraints, but still only represents about 28% of total energy expenditure. Other areas, like manufacturing, healthcare, and communication technology, will be affected in much the same way as transportation, and will have to move toward less total consumption and more efficient machines and system design. How many data centers can society maintain in a demi- or post-Anthropocene scenario? How many hospital ventilators per 100,000 people? Further research and planning in this area remains essential. A second recommendation is to model the intentional slowing of transportation systems and to build the possibility of slower, lower energy-expenditure operation into system design. A third recommendation is to plan for and construct the infrastructure needed for low-energy forms of transportation. A fourth recommendation is to take resources and physical space from high-energy systems of transportation to do so. (Several of these recommendations will be familiar to city planners and urbanists.) A fifth recommendation is to plan for and construct denser human settlements. In addition to sparing energy and carbon emissions, density may be demanded by energy constraints in transportation. How would your city change if no one could commute more than ten kilometers in any direction? A final normative implication regards the necessity of planning for and investing in simpler, physically lighter systems to replace or supplement existing systems of transportation. These systems require less energy to manufacture, maintain, run, and repair, and are less likely to suffer breakdowns in the first place. They are therefore more suited to an energy-constrained world.

It is too much to expect that the downward pressure on transportation speed exerted by decreased energy availability will *reverse* the changes set into physical space by two centuries of faster and more energy-intensive transportation. Future human settlements are unlikely to resemble village idylls, because (i) history is not a time-reversible process and (ii) low-power technologies (e.g., information technology, ball bearings, pneumatic tires) that were limited or non-existent in the pre-velocity age will coexist with lower-energy forms of transportation in the post-velocity age. If the systems for knowledge transmission are more resistant to change and retrenchment than the systems for energy production (which is not certain), humans may pass into a state of relatively high knowledge paired with relatively low energy availability. Technologies that demand a moderate level of knowledge to design and build but a low level of energy to run, like the bicycle, are likely to play an increased role in such societies. The technologies that remain viable in a low-energy world can be expected to create new configurations of social and physical space. And, as documented by Kidder (2009), Oldenziel and de la Bruhèze (2011), and others, new configurations of space loop back into the people who traverse that space, transforming attitudes and behaviors which further alter the built environment [41,42]. One can imagine, for example, suburbs centered on bicycle transportation, because no other mode can effectively reach city centers from outlying residential communities. Just as the car, combined with American urban planning, developed and spread specific forms of human settlement and arrangements of energy across the planet in the twentieth century, the modes of transportation demanded by an energy-constrained system are likely to grow their own topologies of movement and exchange.

10. Conclusions

It is impossible to know the specific effects, challenges, and constraints that will accompany any exit from the human-dominated geological epoch. Most plausible post-Anthropocene scenarios, however, envision a decline (slowly or not) in human control of the material and surface area, and therefore the energy stocks and flows, of Earth. This potential reduction in the total amount of energy available to human societies will constrain the ability of these societies to move people through physical space, a task whose scale and complexity in existing societies remains significant.

This article finds that the choice among different modes of transportation turns out to be non-trivial in energy-constrained scenarios. Energy-efficient means of transportation allow for physically larger human settlements, for example, that can (holding density constant) fit more people than can settlements that rely on less-efficient means of transportation. Bicycles allow for quite large coverage areas, rapid transit for moderately large ones, buses for smaller areas, and cars for coverage areas so small they would be better served by other modes of transportation to begin with. Walking is not meaningfully energy constrained but is constrained by considerations of time, terrain, fatigue, and cargo. Because any post-Anthropocene scenario is likely to involve changes in the structures that preserve and communicate knowledge as well as those that produce and distribute energy, the transportation modes available to future societies will be different, the choices among them will be influenced by new incentives, and their development may rely on previously-developed knowledge more suited to a low-energy world. Finally, transportation after the loss of current levels of energy availability is likely to be slower than existing transportation, even holding mode choice constant, because of the cubic relationship between velocity and the power needed to overcome drag.

Funding: This research received no external funding.

Acknowledgments: The author thanks C. Marino, T. Stringer, M. Toyoshima, and two anonymous reviewers.

Conflicts of Interest: The author declares no conflict of interest.

References

1. Watson, M.J.; Watson, D.M. Post-Anthropocene Conservation. *Trends Ecol. Evol.* **2020**, *35*, 1–3. [CrossRef] [PubMed]
2. Tainter, J.A.; Allen, T.F.H.; Little, A.; Hoekstra, T.W. Resource Transitions and Energy Gain: Contexts of Organization. *CE* **2003**, *7*, 4. [CrossRef]
3. Hall, C.; Balogh, S.; Murphy, D. What is the Minimum EROI that a Sustainable Society Must Have? *Energies* **2009**, *2*, 25–47. [CrossRef]
4. Tainter, J.A. *The Collapse of Complex Societies (New Studies in Archaeology)*; Cambridge Univ. Press: Cambridge, UK, 2011; ISBN 978-0-521-38673-9.
5. Heikkurinen, P. (Ed.) *Sustainability and Peaceful Coexistence for the Anthropocene*; Transnational Law and Governance-Routledge: London, UK; New York, NY, USA, 2017; ISBN 978-1-138-63427-5.
6. Heikkurinen, P.; Ruuska, T.; Wilén, K.; Ulvila, M. The Anthropocene exit: Reconciling discursive tensions on the new geological epoch. *Ecol. Econ.* **2019**, *164*, 106369. [CrossRef]

7. Steffen, W.; Rockström, J.; Richardson, K.; Lenton, T.M.; Folke, C.; Liverman, D.; Summerhayes, C.P.; Barnosky, A.D.; Cornell, S.E.; Crucifix, M.; et al. Trajectories of the Earth System in the Anthropocene. *Proc. Natl. Acad. Sci. USA* **2018**, *115*, 8252–8259. [CrossRef] [PubMed]
8. EIA. How Much Energy Does a Person Use in a Year? Available online: https://www.eia.gov/tools/faqs/faq.php?id=85&t=1 (accessed on 29 January 2020).
9. EIA. Primary Energy Consumption. Available online: https://www.eia.gov/tools/glossary/index.php?id=Primary%20energy%20consumption (accessed on 29 January 2020).
10. Jacobson, M.Z.; Delucchi, M.A.; Bauer, Z.A.F.; Goodman, S.C.; Chapman, W.E.; Cameron, M.A.; Bozonnat, C.; Chobadi, L.; Clonts, H.A.; Enevoldsen, P.; et al. 100% Clean and Renewable Wind, Water, and Sunlight All-Sector Energy Roadmaps for 139 Countries of the World. *Joule* **2017**, *1*, 108–121. [CrossRef]
11. Hansen, K.; Breyer, C.; Lund, H. Status and perspectives on 100% renewable energy systems. *Energy* **2019**, *175*, 471–480. [CrossRef]
12. US Bureau of Transportation Statistics. U.S. Energy Consumption by the Transportation Sector. Available online: https://www.bts.gov/content/us-energy-consumption-transportation-sector (accessed on 30 January 2020).
13. FHWA. Average Annual Miles per Driver by Age Group. Available online: https://www.fhwa.dot.gov/ohim/onh00/bar8.htm (accessed on 30 January 2020).
14. Prokopiak, L. Cost Competitive Electric Vehicle Fleet. Available online: http://large.stanford.edu/courses/2012/ph240/prokopiak1/ (accessed on 30 January 2020).
15. US Bureau of Transportation Statistics. Commute Mode Share. Available online: https://www.bts.gov/content/commute-mode-share-2015 (accessed on 30 January 2020).
16. US Bureau of Transportation Statistics. Energy Intensity of Passenger Modes. Available online: https://www.bts.gov/content/energy-intensity-passenger-modes (accessed on 30 January 2020).
17. Coley, D. *Emission Factors for Walking and Cycling*; Center for Energy and the Environment: Exeter, UK, 2001.
18. Drivers' Annual Mileage Rates Fall. *BBC News*, 29 July 2014.
19. Cox, W. Average Chinese Car Travels as Much as American Car. Available online: https://www.newgeography.com/content/006420-average-chinese-car-travels-much-american-car (accessed on 13 March 2020).
20. Liu, H.; Man, H.; Cui, H.; Wang, Y.; Deng, F.; Wang, Y.; Yang, X.; Xiao, Q.; Zhang, Q.; Ding, Y.; et al. An updated emission inventory of vehicular VOCs and IVOCs in China. *Atmos. Chem. Phys.* **2017**, *17*, 12709–12724. [CrossRef]
21. EPA. Fuel Economy of the 2018 Nissan Leaf. Available online: https://www.fueleconomy.gov/feg/noframes/39860.shtml (accessed on 30 January 2020).
22. Delucchi, M.; Burke, A.; Lipman, T.; Miller, M. Electric and Gasoline Vehicle Lifecycle Cost and Energy-Use Model. Available online: https://escholarship.org/uc/item/1np1h2zp (accessed on 29 January 2020).
23. Stodolsky, F.; Vyas, A.; Cuenca, R.; Gaines, L. *Life-Cycle Energy Savings Potential from Aluminum-Intensive Vehicles*; Argonne National Lab: Lemont, IL, USA, 1995.
24. Łebkowski, A. Studies of Energy Consumption by a City Bus Powered by a Hybrid Energy Storage System in Variable Road Conditions. *Energies* **2019**, *12*, 951. [CrossRef]
25. MacKay, D. *Sustainable Energy—without the Hot Air*; UIT Cambridge: Cambridge, UK, 2010; ISBN 978-0-9544529-3-3.
26. Catling, D.T. Paper 8: Principles and Practice of Train Performance Applied to London Transport's Victoria Line. *Proc. Inst. Mech. Eng. Conf. Proc.* **1966**, *181*, 48–61. [CrossRef]
27. Cushman-Roisin, B.; Tanaka Cremonini, B. Useful Numbers for Environmental Studies and Meaningful Comparisons. Available online: http://www.dartmouth.edu/~{}cushman/books/Numbers/Front-Matter.pdf (accessed on 13 March 2020).
28. Rubin, T. Los Angeles Metro Bus System Compares Favorably with Its Peer Group. Available online: https://www.newgeography.com/content/002361-los-angeles-metro-bus-system-compares-favorably-with-its-peer-group (accessed on 13 March 2020).
29. Graurs, I.; Laizans, A.; Rajeckis, P.; Rubenis, A. *Public Bus Energy Consumption Investigation for Transition to Electric Power and Semi-Dynamic Charging*; Latvia University of Agriculture: Jelgava, Latvia, 2015.
30. Pivit, R. Measuring Aerodynamic Drag. *Radfahren* **1990**, *2*, 44–46.
31. McArdle, W.D.; Katch, F.I.; Katch, V.L. *Essentials of Exercise Physiology*, 2nd ed.; Lippincott Williams & Wilkins: Philadelphia, PA, USA, 2000; ISBN 978-0-683-30507-4.

32. Clark, M.A.; Springmann, M.; Hill, J.; Tilman, D. Multiple health and environmental impacts of foods. *Proc. Natl. Acad. Sci. USA* **2019**, *116*, 23357–23362. [CrossRef] [PubMed]
33. Roche, J. *Agribusiness: An International Perspective*; Routledge: Philadelphia, PA, USA, 2020; ISBN 978-1-138-48865-6.
34. Illich, I. *Energy and Equity*; Marion Boyars: London, UK; New York, NY, USA, 2009; ISBN 978-0-7145-1058-3.
35. Geotab. Gridlocked Cities. Available online: https://www.geotab.com/gridlocked-cities/ (accessed on 29 January 2020).
36. Comiso, J.C.; Meier, W.N.; Gersten, R. Variability and trends in the Arctic Sea ice cover: Results from different techniques: trends in the arctic sea ice cover. *J. Geophys. Res. Oceans* **2017**, *122*, 6883–6900. [CrossRef]
37. SCOR Biological Observatories Workshop. *Report of the Ocean Acidification and Oxygen Working Group*; SCOR: Venice, Italy, 2009.
38. Hall-Spencer, J.M.; Rodolfo-Metalpa, R.; Martin, S.; Ransome, E.; Fine, M.; Turner, S.M.; Rowley, S.J.; Tedesco, D.; Buia, M.-C. Volcanic carbon dioxide vents show ecosystem effects of ocean acidification. *Nature* **2008**, *454*, 96–99. [CrossRef] [PubMed]
39. NASA. Carbon Dioxide. Available online: https://climate.nasa.gov/vital-signs/carbon-dioxide/ (accessed on 29 January 2020).
40. Tisdell, C.A. Economic Benefits and Drawbacks of Cities and Their Growth. Available online: https://ageconsearch.umn.edu/record/90615/ (accessed on 13 March 2020).
41. Kidder, J.L. Appropriating the city: space, theory, and bike messengers. *Theor. Soc.* **2009**, *38*, 307–328. [CrossRef]
42. Oldenziel, R.; Albert de la Bruhèze, A. Contested Spaces. *Transfers* **2011**, *1*, 29–49. [CrossRef]

Article

Envisioning Tourism and Proximity after the Anthropocene

Outi Rantala [1,*], Tarja Salmela [1], Anu Valtonen [2] and Emily Höckert [1]

1 Multidimensional Tourism Institute, University of Lapland, 96300 Rovaniemi, Finland; tarja.salmela@ulapland.fi (T.S.); emily.hockert@ulapland.fi (E.H.)
2 Faculty of Social Sciences, University of Lapland, 96300 Rovaniemi, Finland; anu.valtonen@ulapland.fi
* Correspondence: outi.rantala@ulapland.fi

Received: 24 March 2020; Accepted: 4 May 2020; Published: 12 May 2020

Abstract: The current Earthly crisis demands new imaginings, conceptualisations and practices of tourism. This paper develops a post-anthropocentric approach to envisioning the possibilities of the 'proximate' in tourism settings. The existing generic definitions of proximity tourism refer to a form of tourism that emphasises local destinations, short distances and lower-carbon modes of transport, as well as the mundane exceptionality of the ordinary. We conceptualise proximity tourism with feminist new materialist literature, which accords agency to the ongoing common worlding of all matter—including but not limited to humans—rather than to separate individual agents. More specifically, our research explores the idea of proximity by drawing closer to the geo—to the Earth—through geological walks in the Pyhä National Park in Finnish Lapland. We analyse these walks with the notions of rhythmicity, vitality and care—ideas constructed from the theoretical heritage guiding our study. By doing this, we explore the potential of proximity tourism in ways that intertwine non-living and living matter, science stories, history, local communities and tourism. The outcome of this analysis, we propose, composes one possible narrative of tourism after the Anthropocene.

Keywords: proximity tourism; Anthropocene; more-than-human; new materialism

---

1. Introduction

A great part of the academic debate surrounding the Anthropocene and tourism has focused on climate change. Much of the research on the topic has been dedicated to quantitative studies that predict a substantial acceleration of international tourism on macro levels [1,2]. Tourism researchers Stefan Gössling, C. Michael Hall and David Scott, for example, place a focus on the limitations of technological advancement in compensating for the emissions caused by the rapid growth of tourism, the difficulties of demanding responses to climate change from broader tourist populations, the limited progress towards decarbonisation in tourism and the lack of consensus among tourism leaders on how to contribute to the mitigation of environmental impact [3–6]. Moreover, it has been pointed out that developing countries, which could benefit the most from the contributions of tourism, are also the most vulnerable to climate change, creating a vicious cycle of growing tourism, emissions and vulnerability [7]. As argued by other scholars, the current discourse on sustainability does not provide adequate tools for addressing and solving the profound earthly crisis [8,9] or making change within the tourism sector [2,10]. This is because the associated concepts and practices, such as green growth, still rely on business-oriented thinking driven by the imperative of growth, are human-centric in their scope and perpetuate the nature–culture divide. Therefore, a more radical change in imagining, conceptualising and practising tourism is needed.

In the paper at hand, we propose proximity tourism—and one particular form, geotourism [11–13]—as a possible source of hope for surviving the earthly crisis. Proximity tourism emphasises local destinations, short distances and lower-carbon modes of transportation [14,15]. In a situation where carbon emissions need to be cut down radically, it offers a new way of understanding and orienting oneself towards tourism. In the existing literature, proximity tourism has been approached through questions of attractiveness [16], cultural and physical distance [17], walkability [18] and transportation and accessibility [19]. In the Nordic context, much attention has also been paid to local tourism in second homes [20]. Moreover, Steven Hollenhorst, Susan Houge-Mackenzie and David Ostergren [14] introduce the term locavism, which refers to bioregional tourism that takes place close to home, shifting attention away from distant, exotic places to one's own backyard and favouring slow travel on the ground. The authors argue that "a key element in the shift from tourism to locavism may lie in the realisation that a simple connection to one's human and ecological community is equally valuable and rewarding as distant tourism experiences" [14] (p. 314).

We approach proximity tourism with an example of geotourism, which is, at its best, grounded in sustainable, responsible educational choices and practices, placing a strong emphasis on geoethics and geoconservation [11,13,21]. Thus, our empirical example of proximity tourism—the geo walk—connects our paper with existing research on geotourism. Geotourism—tourism with a geological purpose [11] and the aim to promote awareness of geoheritage [13,22]—connects with Hollenhorst et al. [14]'s community-oriented notions of locavism by identifying the value of human and nonhuman local communities in tourism practice. Geotourism also holds educational potential, motivating people to understand the connections between geological processes and current environmental issues, such as climate change and threats to nature's biodiversity [13] (pp. 4–5). Altogether, geotourism bridges the nature–culture divide and enhance awareness of geoethics [13,23]. Peppoloni and colleagues [23] (p. 31) remind us that *geo* refers to *'gaia'*, meaning 'Earth' (Greek), or 'home, the dwelling place', *'ga'*, based on its Sumerian base. Thus, we treat human interactions with the *geo*, including geotourism practices, as essentially ways of encountering-with [24] and becoming-with the Earth [25]. To become-with the Earth invites us not only to observe from distance, but to dwell—to return *home*. Geological walks are a way to come back and become attuned to our shared histories with and from the Earth.

Continuing down the path opened by the aforementioned seminal research on proximity tourism, we see a need to further deepen its theoretical premises. Our main concern is that the current body of work relies mainly on anthropocentric accounts of travelling close. It pays heightened attention to humans' motivations to practice, or not to practice, proximity tourism, and to the accessibility of proximity tourism destinations. In their study, Hollenhorst et al. [14] link proximity to connection with one's human and ecological community, while Inma Díaz Soria and Joan Carles Llurdés Coit [26] connect it to the appreciation of the mundane exceptionality of the supposedly ordinary—which, we might add, includes the exceptionality of proximate nature and its histories. Proximity could also mean something other than being physically close [26]. Inspired by these unfinished stories of 'proximity', our contribution lies in its exploration of different ways of thinking, doing and researching proximity beyond, but in relation to, the Anthropocene.

In order to experience and understand proximity beyond the mere human community, we apply post-anthropocentric theorising and employ feminist new materialist concepts [27–29]. We understand post-anthropocentric theorising as a feminist undertaking directed towards overcoming species boundaries and violent hierarchies among earthly inhabitants. In Donna Haraway's words [28], it is about crafting kin-stories with multiple Others. This necessitates the decentring of humans as 'masters of the Earth', and instead cherishes the entanglement of all life [27,30] in the past, present and future [31]. Engaging with and becoming responsive to the temporalities, histories and togetherness of all life coincides with 'a return to the Earth' that sets the ground for geotourism. For far too long, the Earth has been considered "too earthy to be worthy of serious attention", being beneath us and not above, in the sky, enabling humankind to look into the heavens and stars [32] (p. 70).

By embracing this earthiness, we consider geotourism as having the innate potential to fuel post-anthropocentric theorising, allowing the conceptualising of tourism beyond the Anthropocene. To drive this change, we employ three theoretical concepts inspired by feminist new materialist heritage—rhythmicity [33], vitality [34] and care [29]—that connect us to Earthly stories and allow us to analyse our empirical example, the geo walk. Methodologically, we lean on the walking-with method [35], which likewise draws from feminist new materialist theory and pays particular attention to the ways in which walkers, as embodied and emplaced within a specific setting, walk with multiple others [36]. Based on this framework, our attention in this paper is placed on the situated, down-to-earth perspective of practising proximity tourism while walking with, and on, the famous and unique geologic formations of our empirical site, the Pyhä National Park in the Finnish Lapland, for which guided geo walks are organized annually during the autumn season.

With these three concepts and our empirical example, we set out to illustrate how we can learn about the complex processes of life constantly taking place proximate to us through our touristic experiences. We explore the ways we can sensitise ourselves both to the new stories that are being born every moment and to the new histories that are being created. Thus, our task is to go beyond the Anthropocene in theorising proximity tourism.

2. Theoretical Orientation: Post-Anthropocentric Theorising

> "Nature can no longer be imagined as a pliable resource for industrial production or social construction. Nature is agentic—it acts, and those actions have consequences for both the human and nonhuman world". [37] (pp. 4–5)

Post-anthropocentric theorising can be traced back to an origin in feminism(s) as an emancipatory theory and philosophy [38] (p. 17). We will first briefly introduce the feminist legacy guiding our paper, after which we will point out the key tenets of feminist new materialism that characterise our post-anthropocentric theorising. These rest on the three concepts guiding our theoretical work, introduced in detail in dialogue with our empirical example of geo walks.

It has been made evident that feminist approaches "have the potential to mount a radical challenge to humanist academic discourses and practices surrounding sustainability, social responsibility and justice" [38] (p. 167, referring to Plumwood, 1993). The possibilities and alternatives provided by feminist theories to thought about the Anthropocene are varied, as exemplified by the valuable contributions of influential feminist scholars in the book *Anthropocene Feminism* [39]. Grusin notes that the concept of the Anthropocene has been present in feminism and queer theory for decades: It is "a genealogy that is largely ignored, or, worse, erased, by the masculine authority of an institutional scientific discourse that now seeks to name our current historical moment the Anthropocene" [39] (p. viii). We argue that the need for new, more complex and sensitive conceptualisations of the Anthropocene suggested by feminist scholars, together with imaginaries of what happens after the Anthropocene, are tightly connected to the study and practice of tourism, as well as to explorations of the role of the post-anthropocentric tourist. Feminist theories disrupt and complicate generic notions of what counts as 'sustainable', 'ethical', 'right' or 'wrong', thus rejecting normative accounts of 'sustainable tourism' that are unavoidably built upon unequal power relations when envisioned from anthropocentric standpoints.

Within feminist theories, feminist new materialism forms the particular inspiration for our study of proximity. In recent decades, feminist new materialism has become a popular approach in various disciplines, ranging from the social sciences, arts, cultural and media studies, science and technology to contemporary philosophy [40] (p. 297). Deconstructing the material/discursive dichotomy "that retains both elements without privileging either" [37] (p. 6), feminist new materialism's practitioners are multidisciplinary and draw from various theoretical heritages, as an interest in matter is not bounded to any academic or scientific discipline [37] (pp. 9–10). However, a focus on 'matter', meaning "a dynamic and shifting entanglement of relations" [30] (p. 224), is a key element in drawing scholars of feminist new materialism together. In their view, matter is simultaneously a verb and a noun: Matter is mattering, an ongoing movement "between nature and culture, the animated and automated, bodies and environments" [41]. As such, its character is "fundamentally multiple, self-organising, dynamic and inventive" [41], fuelling material feminism's demand for "profound—even startling—reconceptualisations of nature" [37] (p. 5).

Mattering, as an ongoing process 'in-between', holds in itself another central tenet of feminist new materialism: Entanglement. Karen Barad notes how we "lack an independent, self-contained existence" [30] (p. ix), an idea that forms the basis for post-anthropocentric theorising. Anna Tsing beautifully describes the meaning of entanglement: " ... rather than limit our analyses to one creature at a time (including humans), or even one relationship, if we want to know what makes places liveable, we should be studying polyphonic assemblages, gatherings of ways of being" [42] (p. 157). In the field of geology, Marcia Bjornerud relates to Tsing's thoughts of polyphony when referring to "our extended family of living organisms", and to the utmost necessity of recognising our place in time in relation to this extended family [43] (pp. 16–17). Geology's acknowledgement of 'deep time'—a timescale that points to the Earth's multi-billion-year history, escaping human comprehension—communicates with the philosophy of mattering. However, as Frodeman [32] (p. 71) notes, there has been a wide neglect of the concept of geologic time within human sciences, pointing to its undiscovered potential. For post-anthropocentric theorising, deep, more-than-human timescales, mattering and entanglement provide ways to undo dichotomies between nature and culture, the discursive and material, the theoretical and empirical and the human and non-human by examining the becoming of these diverse dimensions of the world in relation to each other [44] (p. 28). In this relationality, situatedness becomes important: By acknowledging the liveliness of the material and non-human dimensions of the world, and the productiveness of relations with and among them, feminist new materialist approaches take into account situated knowing in new forms [44] (p. 36).

With this feminist and new materialist legacy, we propose that envisioning ecologically attentive ways of knowing, doing and theorising tourism, with a focus on the 'proximate', is made possible by post-anthropocentric theorising. Fundamentally, such theorising moves beyond human centrism. It requires us to 'notice', meaning that "We need to know the histories humans have made in these places and the histories of nonhuman participants" [42] (p. 160). Post-anthropocentric theorising encourages curiosity towards life beyond, and entangled with, humans. The re-discovery of such curiosity and a related sense of wonder is also considered essential to the development of modern geotourism [13] (p. 8).

Post-anthropocentric theorising is thus a posthumanist project, at the heart of which lies the questioning of the centrality of human power [38] (p. 168). When learning about the complex processes of life that constantly take place outside the everyday lives of humans—the new stories born every moment, the new histories created—it seems outlandish to consider the stories of humans to be the most important or interesting. Yet, as Tsing notes, " ... we are not used to reading stories without human heroes", as human centrism holds tightly to "dreams of progress" [42] (p. 155). This fixation on human protagonists is communicated also in Bjornerud's [43] account on timeliness. Bjornerud points out that we so often lack "a sense of temporal proportion—the durations of the great chapters in Earth's history, the rates of change during previous intervals of environmental instability, the intrinsic timescales of 'natural capital' like groundwater systems" (p. 7). A disinterest towards natural history that characterizes a considerable portion of humans [43] (p. 7) makes post-anthropocentric theorising ever more important. Not being used to something, or lacking an interest in something, does not mean that we cannot learn or become interested. We argue that to take on this task, to go beyond the Anthropocene in theory and practice, is a crucial question of 'response-ability'. Posthuman theorising requires that we become response-able to those we meet and have met [28,45] (p. 130)—and, to add, also to those we have never had a chance to physically encounter, but of whom we can hear stories about—forming a moral obligation to them [46] (p. 16). Through this response-ability, post-anthropocentric theorising becomes material, fleshy, proximate and embedded in situated knowings and histories, instead of something abstract and distant. It is grounded in matter that matters and holds agency "in the world's becoming, in its ongoing 'intra-activity'" 27] (p. 803).

With this focus and emphasis on matter, we have chosen three concepts that guide us in our post-anthropocentric theorising of proximity and proximity tourism. Each of these three concepts—rhythmicity, vitality and care—highlights matter(ing), entanglement, relationality, historicity and situatedness in its own particular way, and as such provides potential analytical approaches to exploring proximity. The selection of these analytical concepts has a story of its own. This story weaves together feminist new materialist theoretical heritage and the very history of our research group. The concepts allow us to analyse, understand and structure proximity and geotourism in ways that have not existed until this day in the field of tourism research. The concepts work thus both as philosophical guidelines and as practical anchors, encouraging us to pause and listen. They also challenge us in understanding the tourism experience under scrutiny: Rocks and stones, which are generally considered inanimate, gain new life when approached, for example, through the concept of vitality. This challenges us to widen our understanding of what vitality genuinely means and how vitality becomes organised temporally and collectively.

Rhythmicity allows us to highlight how more-than-human co-living is conditioned by a range of cosmic and technological rhythms [47]; proximity tourism assumes a different, slower rhythm than does mass tourism [48–50], while *geo*, as a basis for geotourism, brings us back to "our deep roots and permanent entanglement with Earth's history" [43] (p. 16). Rhythmicity makes us think about our place in time, and geology helps us, following Bjornerud, to 'fathom' deep time, which is "arguably geology's single greatest contribution to humanity" [43] (p. 16). Post-anthropocentric theorising also entails an understanding according to which non-humans are attributed with *vitality* and agency [31,37,46,51]. This draws attention to the self-organising vitality of all living systems and helps, as Rosi Braidotti [34] suggests, to disrupt the prevailing hierarchy between earthly beings, providing a way to decentre the human. The notion of *care*, in turn, directs us to consider the ethical relation between human and non-human agents. As Maria Puig de la Bellacasa [29] aptly notes: "Care is a human trouble, but this does not make of care a human-only matter" (p. 2), directing us to notice multiple forms of caring relations. These relations are not conceived through normative moralistic visions of care, understood rather as always open and situated [46,52,53].

3. Method

In our discussion of these three core theoretical concepts, we lean on the walking-with method [35,54] and apply it to an example of a geo walk, arranged in the Pyhä-Luosto National Park in the Finnish Lapland. The national park constitutes a mundane surrounding for our research group, as the national park is situated approximately 150 kilometres from our hometown. We usually visit the park for personal and work-related day trips, staying at the university's base, which is situated at the border of the national park, when we organise research seminars and field courses at the park. Therefore, the park forms for us a fruitful surrounding to explore different ways of thinking, doing and researching proximity. Here, we have chosen to concentrate on the example of the geo walk because it enables us to become attuned to our shared histories with and from the Earth, as we discussed in the introduction.

When producing empirical materials related to the geo walk, we have proceeded in line with the core ideas of the walking-with method. Walking-with is movement-with—movement that invites rhythmic, temporal and affective dimensions into our social-scientific inquiry; that is, dimensions related to our core concepts. Thus, the method enables us to highlight sensuous and rhythmic interrelations [55] (p. 183) and ways of becoming attentive to the ordinariness of our surroundings while walking [35] (p. 20). Central to our methodological approach is the practice of sensitising, thinking-with and being open to the ways that more-than-human entanglements manifest in the particular context in which the geo walk takes place [56,57].

In practice, two of the authors participated in a geo walk in in September 2019. The walk was organised by the local visitor centre and was guided by an experienced geologist. It began with a one-hour lecture explaining the geological history of the area and then proceeded through the park for some eight kilometres, following the trails. The authors walked with the other participants, chatted with them and with the guide, made observations, ate snacks and took photographs. They touched rocks and stones, letting their bodies be affected by them, and felt the slippery wooden duckboards or rocky trails under their boots. While driving back home, they shared their experiences, and they wrote out a research diary the following day. They also linked their own embodied and emplaced experiences with emerging geo-social literature [58,59] that considers the various ways the geologic and social intertwine.

In what follows, we use the empirical insights gained during the walk to reflect on and illustrate what proximity tourism after the Anthropocene could mean and how it could offer a way to respond to the current growth-oriented paradigm underlying tourism. We analyse the geo walk through the notions of rhythmicity, vitality and care, and in the analytical sections that follow, we start with a narrative from the geo walk and follow it with theoretical reflection, in which we entwine new materialist literature with our empirical insights.

4. Rhythmicity

4.1. Empirical Insights

We were sitting in the auditorium of the Naava visitor centre of Pyhä-Luosto National Park with some 20 other people, participating in the geologic walk. We had arrived a bit late, as we were in a hurry again. We took a breath to calm our bodies down and focused on listening to Peter, an experienced geologist, talk about the geological history of the region and the Earth. It soon became clear that our human species, which has enjoyed a very short existence on the geological timeline, does not play much of a role at the geological scale. A good reminder to us, habituated to think of history as human history. Peter's figures and diagrams threw us into deep time: Millions and billions of years run in front of our eyes as geological processes, episodes, epochs, eras, travelling lithospheric plates, ice ages and more followed each other. The rhythms of deep time and the busy, minute-based schedules of our everyday lives entangled and clashed in our bodies. We felt a bit troubled, if not amused; our habituated rhythms were suddenly disrupted.

During the lecture, we gradually began to grasp what had happened before the 35 km long range of hills and fells took the forms that we now know and experience. The fells are the remains of ancient mountains that eroded and shrunk via geological processes over about two billion years, contributing to the appearance of various rock types and the formation of gorges in the park. During the late ice age—not so long ago by geological standards—the region was covered by three kilometres of ice. That is hard to even imagine. The border of the retreating glacier reached Pyhätunturi Fell around 10,000 years ago, and the melting glacier formed the crust, soil, rocks and fells of the park. The ancient rhythms and movements of various earthly matter started to become more tangible to us.

Today, the park attracts an increasing number of domestic and international travellers. The flow of travellers follows the rhythms defined by institutional arrangements (e.g., vacation periods) and the seasons (e.g., winter seasons and seasonal weather conditions). These rhythms play a part in building and changing the shapes and forms of the fells: The movements of walkers and skiers may change the places of rocks and widen the trails, shifting the landscape. Then again, the very same seasonal and weather-based issues that control the rhythms of the travellers likewise shape the geological landscape. The arctic rhythm of winter and summer seasons with its alteration of snow, cold, sun, melting snow and water erodes the fells continuously. The permafrost, for instance, lifts some rocks up and breaks others down. This process is just so slow that it is hard to notice. It is no wonder, therefore, that it is so common to consider rocks as stable and still. Yet, as Elisabeth Povinelli reminds us, "we think something is enduring because we can't see or don't experience the constant wobbling" [60] (p. 182). The rhythms of geological forces are different from those of humans; still, they entangle in this very geo walk.

After the lecture, we went to walk in the park. Peter's talk made the landscape's geological processes seem alive. We could image the way the melting ice, waterfalls and ancient glacial river broke their way through the mountains, billions of years old, eroding them into the low, round fells—how the volume of the flowing water transferred rocks, shaping their figures, making tremendous noise on their way. The park turned into a place full of events, noise and movement. Simultaneously, our bodies felt still and peaceful. We were attuned to the rhythms of the rocks: Their seemingly still rhythm made us stop and feel the stillness. Now, each time we walk at the bottom of the Isokuru gorge, surrounded by massive amounts of rock, we stop to admire the view and sense the atmosphere. Such an experience may also change the very basic rhythm of human life, that of breathing. 'Breath-taking' is more than an expression.

The walk also sensitised us to noticing different types of rocks. There are not only rocks, but particular rocks. In geological terms, we distinguished quartzite and conglomerate, for instance, and the way the shapes of the rocks tell their story: Round rocks have been rolling longer through the ancient rivers. We also learned to read the story of the lichen on top of the rocks. A rock full of lichen has stayed in the same place for centuries, while a 'clean' rock has been 'recently' moved from somewhere. Perhaps the most significant rocks we saw on our walk were those displaying well-preserved ripple marks—a memento of the waves of a sea situated in the park some 1.3 million years ago. They lured us to touch them. We let our fingers slip over the smooth ripples, feeling the rhythm of ancient waves. We touched the past, and the past touched us. We also sensitised ourselves to seeing and admiring rocks that somehow appeared fun, beautiful or remarkable to us due to their particular shape or size. One looked like a seal, another like a monster.

Finally, the quality and quantity of rocks also gave rhythm to the very practice of walking—the pace and tempo of our walking changed in accordance with the rocks on the ground. Duckboards provided a relatively stable ground for walkers, while rocky stones were difficult, especially for those of us habituated to walking on paved streets. Rocks are uneven; their surfaces may be slippery. Rocks may slow down our walking, or invite us to stop, to stay on the ground, to feel it, either standing or sitting. Walking, in this sense, is walking with the rhythm of rocks.

4.2. Theoretical Reflections

In previous new materialist studies, the entangled nature of human and more-than-human rhythms is brought up, for example, in Olivia Davies' and Kathleen Riach's [33] study on bee-work. They observe that, for hobbyists, bee-work is like a 'multispecies choreography', where the work is in tune with the locations and movements of the bees. They also note that in large-scale apicultures, the inspection process is hurried, disrupting the fluid movement, which illustrates the human-centred, industrial development of bee-work. This contradictory nature of cosmic and capitalist rhythms is analysed especially by Henri Lefebvre [47], who develops a rhythmanalytical approach together with his wife Catherine Régulier. This approach, the 'rhythmanalysis' method, invites us to attune ourselves towards the situated and embodied nature of rhythms.

Rhythmanalysis takes rhythm and the dynamics between time and space as its starting point [61]. The approach concentrates on the interferences between cyclical and linear time [47]. In rhythmanalysis, cyclical time refers to the natural and cosmic rhythms to which humans have been exposed from the beginning of time through to the development of modern civilisation. In modernity, their repetition continues, but instead of natural rhythms, the repetition is based on technology, work and production, which constitute linear repetition [62] (p. 87); [63] (p. 6). Hence, linear repetition may resemble a cyclical rhythm, but it can never become an actual rhythm [47]. The rhythmanalytic approach enables attentiveness to specific aspects of rhythms, such as the multiplicity and uniqueness of particular rhythms, how rhythms unite with one another in everydayness, how they are discordant and how harmony is formed by the innumerable rhythms present in the body [47] (p. 16); [64] (p. 150).

Attuning ourselves to the rhythms of the rocks during the geo walk allowed us to slow down. Slowing down can be seen as a counteraction to hectic everyday life and capitalist rhythms (as well as growth-orientation)—however, it can also be considered an act of allowing our bodies to perceive the interfaces of diverse rhythms in and around us. Thus, slowing down enables us to perceive the nuances of more-than-human agency, and it makes visible how diverse agencies are entangled in our proximate surroundings. Attuning ourselves to the rhythms of rocks also sensitises us to noticing different types of rocks and the different levels of history that are materialised in the landscape. It allows us likewise to become proximate with the particulars surrounding us and to learn the history and present of these particulars. This may also make us strive for common futures. Attuning ourselves to the rhythms of rocks can, for example, help us to perceive how arctic rhythms are materialised in particular rocks, making us strive to better understand the dynamics causing the rhythms and how our own actions impact and entangle with these rhythms, creating new histories. Rhythms lead us to a sense of timefulness [43]—a poly-temporal worldview—including "a feeling for distances and proximities in the geography of deep time" (p. 17). Bjornerud [43] considers this poly-temporal worldview vital to creating a more sustainable future in the era of the Anthropocene. Geo walks provide one opportunity to understand and become sensitive to poly-temporality and timefulness. This focus on rhythms and temporalities can also invite surprising, even reversed, notions of the rhythmic entanglement of rocks and humans; described by Bjornerud in her prologue on timefulness on the Svalbard islands in the Norwegian arctic, the remains of human history on the islands, human-made artefacts, seem to her older and shabbier than the ancient mountains, which are robust and vital [43] (p. 5). Acknowledging this rhythmic interplay is essential when our aim is to reach out for accounted, situated knowings that acknowledge the relationalities of the more-than-human world.

5. Vitality

5.1. Empirical Insights

While walking in the national park, we were compelled to wonder at, and experience, the vitality of the rocks. We, and many others, commonly consider rocks to be stable, inert and passive matters —as reflected in commonplace sayings such as 'rock solid'. Yet, Peter's talk about geological processes concretised what we have been learning while familiarising ourselves with the geo-social literature. Rocks are vital, 'lively'; they evolve, change and move. This liveliness often goes unnoticed in the expansive timeline of geology, as we discussed earlier. While walking, the theoretical idea of the vitality of materiality thus became lively to us. We understood that rocks are lively also in the sense that they are agentic, acting on and with us, agency emerging as the effect of configurations of human and nonhuman forces. Rocks and stones shaped and framed our doings, guiding how and where to walk, where to put our feet while taking steps. Rocks and stones are also agentic in the sense that they have the power to attract visitors, as Pyhä-Luosto National Park exemplifies so well. Last year (2019), 169,700 visitors came to the park (https://www.metsa.fi/kayntimaarat). The world is full of similar examples—the Grand Canyon, for instance, provides a case in point.

While walking behind the group of participants, one of us started talking about the inspiring question raised by anthropologist Hughes Raffles [65]: What can a rock do? A short reflection on our own experiences rendered visible the fact that rocks and stones can do a lot. For example, they can heal, as in the case of stone therapy. In other words, rocks affect. The very nature of the body means that it is continuously affecting and affected by other bodies—including non-human bodies, since "organic and inorganic bodies all are affective", as Bennett [66] (p. xii) reminds us. When relating to the rocks, a body's agentic capacity may be changed—strengthened, impressed, effectuating perhaps a change of mood. We experienced this change while walking with the rocks; they (and our entire surroundings) made us relax, as they have done so many times before. We felt, indeed, livelier, more vital.

We continued wondering how the very vitality of this particular region we were visiting is largely dependent on these rocks. The fells and rocks, and the preceding geological processes, make touristic activities possible in the first place. The livelihood of the region has rocky roots, so to speak. As a result, we can now go to a restaurant for dinner, ski on the tracks or go downhill skiing, climb with a professional guide, and so on. Many local people derive their living from the rocks, indirectly or directly, like those working in the nearby Lampivaara amethyst mine. Besides providing livelihood for humans, these rocks and stones enable many different types of lichens to live and provide nests for snakes, insects and some birds. Going further, all of life, and all earthlings, are entirely dependent on rocks, stones and minerals. This dependency includes our affluent way of life: Even the mobile phone with which we take photos while walking consists of some thirty minerals. This might open up a space for considering rocks and stones with appreciation and care. As Bennett notes: "The ethical task at hand is to cultivate the ability to discern nonhuman vitality, to become perpetually open to it" [66] (p. 14). The walk cultivated our abilities to discern the vitality of rocks and to become open to it, making us see surroundings that were familiar to us in a new light.

5.2. Theoretical Reflections

In new materialist literature, the notion of vitality draws attention to the self-organising vitality of all living systems [34]. This shift entails blurring the boundaries among what we in the era of the Anthropocene might consider living, semi-living and non-living [46] (p. 112–113). Braidotti [34] suggests that the concept of vitalist materialism constitutes "the core of a posthuman sensibility that aims at overcoming" (p. 55–56). Her idea of vitality draws inspiration from Baruch Spinoza's notion of a monistic universe, which puts in question dualistic oppositions between matter and humans, nature and culture. Spinoza's monistic worldview aims at creating non-dialectical understandings of materialism itself, and it has enabled further definitions of matter as vital and self-organising. It is these monistic premises of the Spinozist legacy that Braidotti uses as a building block for a posthuman theory that escapes anthropocentrism. In Braidotti's thinking, monism, the unity of all living matter and post-anthropocentrism are connected as a general frame of reference for contemporary subjectivity [34] (p. 57).

Braidotti's vitalist approach to living matter displaces the boundary between the portion of life that has been traditionally reserved for the human and the wider scope of non-human life, also known as *zoe*. Braidotti [34] (p. 60) calls attention to the non-human, dynamic, transversal and vital force of life that reconnects previously segregated species, categories and domains. For her, this life, this zoe, is an inhuman force that stretches beyond life. Indeed, Braidotti does not settle for a drastic restructuring of human relations with animals, but suggests that the post-anthropocentric shift requires a planetary, zoe- and geo-centred perspective—that is, a reconfiguring of our relationship to the complex habitat we used to call 'nature'. By complex habitat, Braidotti [34] (p. 81) refers to the 'milieu' of human and non-human inhabitants of this particular planet. Within Spinoza's monistic framework, this concept means that we are all part of 'nature' [67]. Hence, Braidotti [34] (p. 82) encourages us to envision a geo-centred subject as a transversal entity, an enlarged sense of community, encompassing humans, animals and the earth as a whole. This task requires questioning the hierarchical idea of human exceptionalism and letting go of the need to dominate and control nature. It requires recognising zoe and vitality in places, beings and things that we have overlooked in the past.

In a similar vein, Mick Smith engages with rocks as the earth's continental drifters that constitute the thin lithospheric crust that keeps travelling across the planet, on which all of us earthlings are entirely dependent [31] (p. 165). While our current global economy is busy carving its effect into this crust [31], tourism in the post-Anthropocene would mean travelling and living with the rocks. Perhaps it would mean being in transition, like mountains that "grow and move, flow and shrink and perish—eroded by water and ice, exploding in volcanic ecstasies, melting and slipping back into the torrid heat of the Earth's mantle" [31] (p. 165)? Transition, Bjornerud [43] notes, is something that future geologists are taught to understand as the essence of rocks from the early stages of their education: "Rocks are not nouns but verbs—visible evidence of processes: A volcanic eruption, the accretion of a coral reef, the growth of a mountain belt. Everywhere one looks, rocks bear witness to events that unfolded over long stretches of time" [43] (p. 8). Smith [31] (p. 166) suggests that being proximate with geological remnants, like rocks and fossils, enables us to recall something from the past back into life in the present. That is, it lets past lives matter by 'breathing new life' into them.

6. Care

6.1. Empirical Insights

Our walk took place in a national park managed by Metsähallitus, a state-owned enterprise whose task is to take care of and protect the various earthly creatures inhabiting the park. While walking, we therefore saw several noticeboards that advised us how to move, where to move and what is and is not allowed to be done in the park. There were, for instance, boards that told us that it is forbidden to go to some places during certain periods of time to protect rare flora, and others that told us that making a fire and camping is allowed only in marked places.

Besides these instructions, the material arrangements of the park articulate care. The duckboards, for example, prevented us from harming the fragile nature of the national park; at the same time, they mediated the encounter between our bodies and the rocks. For instance, while walking at the bottom of Isokuru gorge, we stepped on stable wooden planks instead of on the floating stones. The stable planks felt safer under our feet, which were used to walking on asphalt or indoors [68]. As the wooden planks are wide at some parts of the trail, they also enable disabled bodies to visit the park, as well as families with small children in strollers. Different groups of people are thus taken care of. This includes animal companions as well, dogs in particular. We saw many dogs during our walk, and one of the participants let us know that the wooden planks are comfortable for dogs' paws, whereas those made of metal are not—even though they would last longer than wooden ones. We therefore started to wonder whether humans were caring for humans, or for human companions, by covering the difficult parts of the paths to make them more easily accessible for visitors. Is it the human that aims to minimise the erosion of the rocks and to maintain the wellbeing of the rocks, or are the wooden planks caring for the visitors?

Besides institutionalised care in the form of protection, guidelines and prohibitions, we noticed other forms of care—and disruptions of care—while walking through the trail. In particular, we paid attention to the human-made piles of stones that stood next to trails. At first glance, the habit of making a pile of some three to five stones seems rather 'innocent', or even cute; yet, on second thought, it is a violent act. It speaks to the human desire to leave a trace of his or her visit. It also tells us about ignorance of the rules of the park, and of the consequences of the act on the biodiversity surrounding the rocks. The lichen on the surface of a stone takes hundreds of years to grow, and the stone may afford shelter for several tiny creatures. Then, all of a sudden, a human hand moves it away—presumably without any ethical thought. The customer manager of the park that we had met earlier told us that she developed a habit of deconstructing the piles while trekking in the park. "Should we do the same?" we think. The habit of making piles of stones is relatively common among trekkers, and at some fells and mountains, they are important guides, helping trekkers to avoid getting lost. However, here, they have no such function. On the contrary, as the local newspaper Lapin Kansa writes on June 6, 2019: "The piling of stones compares to littering".

During the winter season, the rocks are protected under the snow, we discuss while driving back home; or are they? We know that some like to do back-country skiing in the gorge—even though it is forbidden. The avalanches that skiing may bring about might cause the movement of rocks and thereby change the living conditions of the unique mosses and flowers growing there. So many seem to care merely for the human experience, leaving other creatures and beings unnoticed, we both silently think. At almost the same time, the inspiring quote from Maria Puig de la Bellacasa [29] comes to our minds: "Care is a human trouble, but this does not make of care a human-only matter" (p. 2). Caring relations with earthly creatures matter.

6.2. Theoretical Reflections

Care pushes us to think-with. Care—who am I walking with? Care—whose path am I crossing? Care—whose presence can I sense, and whose can I not? This response-ability cannot be chosen—we become response-able [28]. The question is, how do we act, not merely react, thereafter? Care cannot be described as something normative and easily explained: It is always open and situated [46,52,53]. Puig de la Bellacasa notes how "(c)ertainly any notion that care is a warm pleasant affection or a moralistic feel good attitude is complicated by feminist research and theories about care" [52] (p. 2). What does this mean? Care is open: There are always different viewpoints and elements of care that conflict with one another. Care is situated: There is no universal 'answer' to questions of care, and sometimes care means different things in the same places [52] (p. 1). Moreover, care is ethics. Care is entanglement. Care is more than human. Care is matter. What then do we accomplish with thinking about, and with, care?

For Puig de la Bellacasa, care is significant "for thinking and living in more-than-human worlds" [52] (p. 1). For her, care is speculative ethics (p. 69). Care invites us to acknowledge, and be curious about, our more-than-human world. Living in a more-than-human world means we are, day after day, minute after minute, in the midst of ethically charged encounters, entanglements and clashes with others. Moreover, our relations with others do not only 'involve care'; "care is relational per se": "Caring and relating share ontological resonance" (p. 69). These encounters are care-laden, as "[c]are is omnipresent, even through the effects of its absence" (p. 1). Puig de la Bellacasa notes that "for interdependent beings in more-than-human entanglements, there has to be some form of care going on somewhere in the substrate of their world for living to be possible" (p. 5; see also p. 70). Care is unavoidably, ontologically more than human by nature.

Moreover, feminist philosophy leads us to consider care—always rooted in ethics—as a prerequisite for life in a messy, complex compost of more-than-human relationalities [28,42,52]. Our first consideration of care, relevant for our study, points to the caring aspect of care—the actual, practical acting of/with care. Following Haraway [28], care is about taking responsibility and facing our response-ability with/in our world. Caring—not only in the maternal sense—is becoming response-able with the creatures we encounter [28]. Care is taking responsibility for our actions and encounters with others. Even more so, it demands action, 'maintenance work', to make affectivity. As 'part of situations of care', becoming caring—as an active verb, becoming response-able—is thus more than a moral intention: To 'care about', instead of 'caring for' [52,69] (p. 5).

Institutionalised care as part of the geo walk experience is one form of the 'maintenance work' of care, taking the form of materialised protection, guidelines and prohibitions in the national park. It is the active practising of care. This, however, does not suggest caring ought to be considered an obligation, which would de-naturalise its existence [52] (p. 70). This principle applies even when the maintenance of nature parks is mediated by law and regulations. There is no regulation that obliges one to deconstruct the tourist-made piles of stones in national parks. This form of practising response-ability is situated and non-normative. It happens through the embodied engagement of the member of the staff of the nature park with the more-than-human community inhabiting the nature parks. It is a caring relation that encompasses respect, attentiveness and more-than-human solidarity.

Post-anthropocentric understanding of care thus includes a realisation that care is not a human-only matter, even when it is 'human trouble' [29] (p. 2). A recognition of humans as (only one) part of more-than-human assemblages and thus constitutive only through others allows us to notice multiple forms of caring relations. Caring takes place in various forms, spaces, histories and stories on the more-than-human planet. Trees care for each other, as do bears, birds, butterflies and worms. Humans are part of this caring worlding process: Humans can take care of worms, too. Bees take care of humans while pollenating plants and making the world a liveable place. Not all caring processes are identifiable or rationalisable. However, they make up an ongoing common worlding built upon relationality [28].

Lastly, caring points to proximity and distance. What does it mean to care in close proximity with others? What does it mean to care from a distance? Caring for, as a less mutual and entangled realm of caring than that of becoming response-able with, invites the possibility of caring from a distance. This possibly does not make caring worth any less—but it is different. Proximity invites a different kind of understanding of care. Care comes close: To the encounter, to the intimate, to the everyday. It pushes one to take responsibility. Proximity demands taking action—when proximate, things cannot be put aside, or at least not as easily. Proximity might also demand a slowing down in response-ability, as it did for the authors taking part in geo-walks. Caring with the proximate denies any quick answers or solutions.

7. Conclusions

In his writing on posthumanist tourism, Smith [31] emphasizes the importance of being proximate with non-human others. He uses the example of human encounters with fossils, suggesting that keeping a fossil on one's palm can "help to re-envision links between traces of the past being(s), intensities of present experiences, and future oriented ecological concerns regarding the Anthropocene" [31] (p. 167). According to Smith, these kinds of encounters can offer exemplary openings into our entanglements with the past, present and future in a way that can change our understanding fundamentally. In line with Smith, our aim in this paper has been to illustrate how we can learn about the complex processes of life constantly taking place proximate to us through touristic experiences. We have explored how we can sensitise ourselves both to the new stories that are being born every moment and to the new histories that are being created. We have deliberated how these moments of learning can be identified in a geotourism that intertwines non-living and living matter, science, histories, local communities and tourism. Our task has been to go beyond the Anthropocene in theorising proximity tourism, bearing in mind that posthuman theorising requires becoming response-able with those we meet. As a tool to reach beyond the Anthropocene, we have applied post-anthropocentric theory, and especially the feminist new materialist approach. With this approach, our focus has been on the liveliness of material, the situatedness of knowing and the productiveness of caring relations. Furthermore, we have chosen three concepts that carry new materialist heritage—rhythmicity, vitality and care—to discuss and illustrate potential analytical approaches to proximity tourism. Envisioning tourism in the post-Anthropocene requires both a conceptual creativity and a linguistic clarity that can create new collective imaginaries [34]. Hence, in our discussion of our three concepts, we have used the example of a geo walk in the Finnish Lapland to visualise the imaginaries we refer to. The empirical example of the geo walk has enabled us to experience glimpses of Braidotti's idea of geo-centred, monist subjectivity that encompasses humans, non-humans and the earth as the whole [34] (p. 58).

Our empirical example has highlighted how the concept of rhythmicity makes visible the ways that the past, present and future become alive and entangled in our proximate surroundings. The concept also enables us to see how the rhythms of production and technology intertwine with cosmic rhythms [47], urging us to move beyond a romantic idea of proximity tourism—beyond the idea of proximity tourism as some kind of pure connection with the more-than-human, separate from capitalist and growth-oriented rhythms. Instead, rhythmicity pushes us towards a conceptualisation of proximity in tourism as an entangled and situated way of slowing down in our surroundings and as an opportunity to learn about the particulars of our surroundings—particulars where the various rhythms and agencies intersect and intertwine. Slowing down, attuning, and learning can evoke care instead of resistance and avoidance.

Our idea of proximity tourism also resists the traditions depriving vitality from matter and reducing it to a mere substrate for human re-creation [46]. It calls for the recognition of human participation in a shared, vital materiality [66] (p. 14). As Bennett [66] further argues, "if we continue to see the things as passive objects, it encourages (and legitimises) us to ignore the vitality of matter and the lively powers of material formations. If rocks are only considered as 'resources' or 'threats' to humans (e.g., in earthquakes), then thinking with rocks as vital extends our ethical and political response" [54] (p. 851).

Furthermore, reading the geo walk through the concept of care has invited us to see care as more than a human matter [52]. Care is grounded in ethics and ethico-politics and is present in post-anthropocentric tourism narratives and practices, such as the geo walk. On the geo walk, care manifests, to begin with, in the institutionalised practices of nature parks—practical acts of becoming response-able [28]—as active processes of caring. Moreover, and most importantly, institutionalised practices of care entangle with other more-than-human caring relations, including those that take place without human influence. The concept of care has also invited us to consider proximity as a push to become response-able with the more-than-human planet we inhabit together with multiple others.

Martin Gren [24] asks how the modern human condition, and the moderns' understanding of tourism, inevitably changes when we encounter "the Earth of the Anthropocene". While we have proposed proximity tourism as a vision of tourism after the Anthropocene, it can simultaneously be seen as a way of enhancing the common chances of surviving through the Anthropocene. In sum, we do not propose that proximity tourism should be considered a particular, distinct form of tourism. Instead, proximity tourism ought to be acknowledged as a sensitive way to orient ourselves within our everyday surroundings. Such an orientation considers different temporalities, entangled and vital materiality and the different manifestations of care in a more-than-human world. We thereby suggest that proximity tourism should not be developed in opposition to global mass tourism and capitalistic ideals of growth. Rather, the potential of proximity tourism lies in its dynamic and polymorphous, open-ended nature, making it possible for ideas of proximity to become part of other, existing forms of tourism and tourism discourses. Importantly, proximity tourism—equipped with a conceptualisation of proximity that goes beyond the Anthropocene—has the potential to transform the habituated ways of thinking about and practising tourism from within. It fosters the recognition of our inevitable relatedness with the more-than-human world, and this sensitising, we propose, might provide one possible way of practising tourism.

Moreover, a return to the geo—the Earth—can teach us a considerable amount about proximity. Geotourism is one way to rediscover a sense of wonder at the proximate [13], not least by way of intertwining non-living and living matter, science stories, history, local communities and tourism. Rocks and landscapes start to gain new life—vitality—through stories that can be both scientific and mythical. Cultural associations of the geo bring the geological timespan—deep time—closer to humans grasped by an urban, technologised lifestyle [70]. To return to the geo is to turn to a home that is intensely *ours*, in a more-than-human sense—and thus, it plays a critical role in the imagining of tourism after the Anthropocene.

Author Contributions: All the authors have equally contributed to the development of the core idea and argument of the paper, as well as the theoretical framing. O.R. managed the writing process, and contributed particularly to the positioning of the study, to the development of 'rhythmicity' and to the conclusion section. T.S. was responsible for designing and writing the theoretical section. A.V. was responsible for analysing the empirical materials and writing the empirical vignettes. T.S. also actively participated in writing the section on 'care', while E.H. took responsibility for 'vitality'. They also contributed to drawing the conclusions. All authors have read and agreed to the published version of the manuscript.

Funding: This research was funded by Academy of Finland, grant number 324493.

Conflicts of Interest: The authors declare no conflict of interest.

References

1. Gren, M.; Huijbens, E.H. Tourism and the anthropocene. *Scand. J. Hosp. Tour.* **2014**, *14*, 6–22. [CrossRef]
2. Gren, M.; Huijbens, E. *Tourism and the Anthropocene*; Routledge: London, UK, 2016.
3. Gössling, S.; Hall, C.M.; Peeters, P.; Scott, D. The future of tourism: Can tourism growth and climate policy be reconciled? A mitigation perspective. *Tour. Recreat. Res.* **2010**, *35*, 119–130. [CrossRef]
4. Gössling, S.; Scott, D.; Hall, C.M.; Ceron, J.-P.; Dubois, G. Consumer behaviour and demand response of tourists to climate change. *Ann. Tour. Res.* **2012**, *39*, 36–58. [CrossRef]
5. Scott, D.; Hall, C.M.; Gossling, S. A report on the Paris Climate Change Agreement and its implications for tourism: Why we will always have Paris. *J. Sustain. Tour.* **2016**, *24*, 933–948. [CrossRef]

6. Gössling, S.; Scott, D. The decarbonisation impasse: Global tourism leaders' views on climate change mitigation. *J. Sustain. Tour.* **2018**, *26*, 2071–2086. [CrossRef]
7. Scott, D.; Hall, C.M.; Gössling, S. Global tourism vulnerability to climate change. *Ann. Tour. Res.* **2019**, *77*, 49–61. [CrossRef]
8. Banerjee, S.B. Who sustains whose development? Sustainable development and the reinvention of nature. *Organ. Stud.* **2003**, *24*, 143–180. [CrossRef]
9. Heikkurinen, P. *Sustainability and Peaceful Coexistence for the Anthropocene*; Routledge: London, UK, 2017.
10. Moore, A. *Destination Anthropocene: Science and Tourism in The Bahamas*; University of California Press: Oakland, CA, USA, 2019.
11. Allan, M. Geotourism: An opportunity to enhance geoethics and boost geoheritage appreciation. *Geol. Soc. Lond. Spéc. Publ.* **2015**, *419*, 25–29. [CrossRef]
12. Aquino, R.; Schanzel, H.; Hyde, K.F. Unearthing the geotourism experience: Geotourist perspectives at Mount Pinatubo, Philippines. *Tour. Stud.* **2017**, *18*, 41–62. [CrossRef]
13. Gordon, J.E. Geoheritage, geotourism and the cultural landscape: Enhancing the visitor experience and promoting geoconservation. *Geosciences* **2018**, *8*, 136. [CrossRef]
14. Hollenhorst, S.J.; MacKenzie, S.H.; Ostergren, D.M. The trouble with tourism. *Tour. Recreat. Res.* **2014**, *39*, 305–319. [CrossRef]
15. Jeuring, J.; Diaz-Soria, I. Introduction: Proximity and intraregional aspects of tourism. *Tour. Geogr.* **2016**, *19*, 4–8. [CrossRef]
16. Jeuring, J.; Haarsten, T. The challenge of proximity: The (un)attractiveness of near-home tourism destinations. *Tour. Geogr.* **2017**, *19*, 118–141. [CrossRef]
17. Kastenholz, E. 'Cultural proximity' as a determinant of destination image. *J. Vacat. Mark.* **2010**, *16*, 313–322. [CrossRef]
18. Hall, C.M.; Ram, G.; Shoval, N. *The Routledge International Handbook of Walking*; Routledge: New York, NY, USA, 2018.
19. Peeters, P.; Dubois, G. Tourism travel under climate change mitigation constraints. *J. Transp. Geogr.* **2010**, *18*, 447–457. [CrossRef]
20. Müller, D.; Hoogendoorn, G. Second homes: Curse or blessing? A review 36 years later. *Scand. J. Hosp. Tour.* **2013**, *13*, 353–369. [CrossRef]
21. Fish, J. Living geoparks in an emerging ecological civilization: A constructive postmodern perspective. *Int. J. Geoheritage* **2013**, *1*, 39–53.
22. Martini, G. Geological heritage and geo-tourism. In *Geological Heritage: Its Conservation and Management*; Barettino, D., Wimbledon, W.A.P., Gallego, E., Eds.; Instituto Technológico Geominero de España: Madrid, Spain, 2000; pp. 147–156.
23. Peppoloni, S.; Bilham, N.; Di Capua, G. Contemporary geoethics within the geosciences. In *Exploring Geoethics*; Springer International Publishing: Cham, Switzerland, 2019; pp. 25–70.
24. Gren, M. Mapping the Anthropocene and tourism. In *Tourism and the Anthropocene*; Gren, M., Huijbens, E., Eds.; Routledge: London, UK, 2016; pp. 171–188.
25. Haraway, D. *When Species Meet*; University of Minnesota Press: Minneapolis, MN, USA, 2008.
26. Díaz Soria, I.; Llurdés Coit, J. Thoughts about proximity tourism as a strategy for local development. *Cuad. Tur.* **2013**, *32*, 65–88.
27. Barad, K. Posthumanist performativity: Toward an understanding of how matter comes to matter. *Signs* **2003**, *28*, 801–831. [CrossRef]
28. Haraway, D. *Staying with the trouble. Making kin in the chthulucene*; Duke University Press: Durham, UK; London, UK, 2016.
29. Puig de la Bellacasa, M. *Matters of Care: Speculative Ethics in More than Human Worlds*; University of Minnesota Press: Minnesota, USA, 2017.
30. Barad, K. *Meeting the Universe Halfway: Quantum Physics and the Entanglement of Matter and Meaning*; Duke University Press: Durham, UK; London, UK, 2007.
31. Smith, M. The Anthropocene: The eventual geo-logics of posthuman tourism. In *New Moral Natures in Tourism*; Grimwood, B.S.R., Caton, K., Cooke, L., Eds.; Routledge: London, UK, 2018; pp. 164–180.
32. Frodeman, R. Hermeneutics in the field: The philosophy of geology. In *The Multidimensionality of Hermeneutic Phenomenology. Contributions to Phenomenology*; Springer: Cham, Switzerland, 2014; Volume 70.

33. Davies, O.; Riach, K. From manstream measuring to multispecies sustainability? A gendered reading of bee-ing sustainable. *Gend. Work. Organ.* **2018**, *26*, 246–266. [CrossRef]
34. Braidotti, R. *The Posthuman*; Polity Press: Cambridge, UK, 2013.
35. Springgay, S.; Truman, S.E. *Walking Methodologies in a More-Than-Human World: WalkingLab*; Routledge: London, UK, 2018.
36. Salmela, T.; Valtonen, A. Towards collective ways of knowing in the Anthropocene: Walking-with multiple others. *Matkailututkimus* **2019**, *15*, 18–32. [CrossRef]
37. Alaimo, S.; Hekman, S. *Material Feminism*, Indiana University Press: Bloomington, IN, USA, 2008.
38. Hamilton, L.; Taylor, N. *Ethnography after Humanism. Power, Politics and Methods in Multi-Species Research*; Palgrave Macmillan: London, UK, 2017.
39. Grusin, R. Introduction: Anthropocene feminism. An experiment in collaborative theorizing. In *Anthropocene Feminism*; Grusin, R., Ed.; University of Minnesota Press: Minneapolis, London, UK, 2017; pp. xii–xix.
40. Revelles-Benavente, B.; Ernst, W.; Rogowska-Stangret, M. Feminist new materialisms: Activating ethico-politics through genealogies in social sciences. *Soc. Sci.* **2019**, *8*, 296. [CrossRef]
41. Coleman, R.; Page, T.; Palmer, H. Feminist New Materialist Practice: The Mattering of Methods. Mai: Feminism & Visual Culture 2019, Focus Issue Intro. Available online: https://maifeminism.com/feminist-new-materialisms-the-mattering-of-methods-editors-note/ (accessed on 17 March 2020).
42. Tsing, A.L. *The Mushroom at the End of the World. On the Possibility of Life in Capitalist Ruins*; Princeton University Press: Princeton, NJ, USA, 2015.
43. Bjornerud, M. *Timefulness-How Thinking like a Geologist Can Help to Save the World*; Princeton University Press: Princeton, NJ, USA, 2018.
44. Leppänen, T.; Tiainen, M. Feministisiä uusmaterialismeja paikantamassa: Materian toimijuus etnografisessa taiteen- ja kulttuurintutkimuksessa. *Sukupuolentutkimus-Genusforskning* **2016**, *29*, 27–44.
45. Despret, V. The Body We Care for: Figures of Anthropo-zoo-genesis. *Body Soc.* **2004**, *10*, 111–134. [CrossRef]
46. Zylinska, J. *Minimal ethics for the Anthropocene*; Open Humanities Press: Ann Arbor, MI, USA, 2014.
47. Lefebvre, H. *Rhythmanalysis. Space, Time and Everyday Life*; Continuum: London, UK, 2004. First published in 1992 as *Élements de rythmanalyse*.
48. Rantala, O.; Valtonen, A. A rhythmanalysis of touristic sleep in nature. *Ann. Tour. Res.* **2014**, *47*, 18–30. [CrossRef]
49. Varley, P.J.; Semple, T. Nordic slow adventure: Explorations in time and nature. *Scand. J. Hosp. Tour.* **2015**, *15*, 73–90. [CrossRef]
50. Rantala, O. With the rhythm of nature: Reordering everyday life through holiday living. In *Theories of Practice in Tourism*; Halkier, H.L.J., Ren, C., Eds.; Routledge: London, UK, 2019; pp. 58–76.
51. Barad, K. Getting real: Technoscientific practices and the materialization of reality. *Differ. J. Fem. Cult. Stud.* **1998**, *10*, 87–128.
52. De La Bellacasa, M.P. 'Nothing comes without its world': Thinking with care. *Sociol. Rev.* **2012**, *60*, 197–216. [CrossRef]
53. Pullen, A.; Rhodes, C. Ethics, embodiment and organizations. *Organization* **2014**, *22*, 159–165. [CrossRef]
54. Springgay, S.; Truman, S.E. Stone Walks: Inhuman animacies and queer archives of feeling. *Discourse Stud. Cult. Politi-Educ.* **2016**, *38*, 851–863. [CrossRef]
55. Vannini, P. Non-representational ethnography: New ways of animating lifeworlds. *Cult. Geogr.* **2014**, *22*, 317–327. [CrossRef]
56. Höckert, E.; Rantala, O. A new materialistic reading of proximity tourism. *Tour. Cult. Commun.*. Submitted manuscript.
57. Rantala, O.; Valtonen, A.; Salmela, T. Walking-with rocks: With care. In *Reimagining Ethics and Politics of Space for the Anthropocene*; Valtonen, A., Rantala, O., Farah, P., Eds.; Edward Elgar Publishing: Cheltenham, UK, forthcoming.
58. Clark, N.; Yusoff, K. Geosocial formations and the Anthropocene. *Theory Cult. Soc.* **2017**, *34*, 3–23.
59. Palsson, G.; Swanson, H.A. Down to earth. Geosocialities and geopolitics. *Environ. Humanit.* **2013**, *8*, 149–171. [CrossRef]
60. Povinelli, E.A.; Coleman, M.; Yusoff, K. An Interview with Elizabeth Povinelli: Geontopower, Biopolitics and the Anthropocene. *Theorycult. Soc.* **2017**, *34*, 169–185. [CrossRef]

61. Edensor, T. Introduction. Thinking about rhythm and space. In *Geographies of Rhythm, Nature, Place, Mobilities and Bodies*; Edensor, T., Ed.; Ashgate: Surrey, UK, 2010.
62. Gardiner, M. Critiques of Everyday Life. Routledge: London, UK, 2000.
63. Lefebvre, H. *Everyday Life in the Modern World*; Transaction Publishers: New Brunswick, Canada, 1984.
64. Meyer, K. Rhythms, Streets, Cities. In *Space, Difference, Everyday Life. Reading Henri Lefebvre*; Goonewardena, K., Kipfer, S., Milgrom, R., Schmid, C., Eds.; Routledge: New York, NY, USA, 2008; pp. 147–160.
65. Raffles, H. Twenty-Five Years is a Long Time. *Cult. Anthr.* **2012**, *27*, 526–534. [CrossRef]
66. Bennett, J. *Vibrant Matter. A Political Ecology of Things*; Duke University Press: Durham, UK; London, UK, 2010.
67. Morton, T. *The Ecological Thought*; JSTOR: New York, NY, USA, 2012.
68. Ingold, T. Culture on the Ground. *J. Mater. Cult.* **2004**, *9*, 315–340. [CrossRef]
69. Tronto, J.C. *Moral Boundaries. A Political Argument for and Ethics of Care*; Routledge: London, UK, 1993.
70. Bohle, M.; Sibilla, A.; I Graells, R.C. A concept of society-earth-centric narratives. *Ann. Geophys.* **2017**, *60*. [CrossRef]

Article

The Post-Anthropocene Diet: Navigating Future Diets for Sustainable Food Systems

Rachel Mazac [1,2,*] **and Hanna L. Tuomisto** [1,2]

1 Department of Agricultural Sciences, Faculty of Agriculture and Forestry, University of Helsinki, Yliopistonkatu 4, 00100 Helsinki, Finland; hanna.tuomisto@helsinki.fi
2 Helsinki Institute of Sustainability Sciences (HELSUS), University of Helsinki, Yliopistonkatu 4, 00100 Helsinki, Finland
* Correspondence: rachel.mazac@helsinki.fi; Tel.: +358-29-4157945

Received: 9 January 2020; Accepted: 15 March 2020; Published: 18 March 2020

Abstract: This article examines how future diets could reduce the environmental impacts of food systems, and thus, enable movement into the post-Anthropocene. Such non-anthropocentric diets are proposed to address global food systems challenges inherent in the current geological epoch known as the Anthropocene—a period when human activity is the dominant cause of environmental change. Using non-anthropocentric indigenous worldviews and object-oriented ecosophy, the article discusses changes in ontologies around diets to consider choices made in the present for sustainable future food systems. This article conceptually addresses, how can pre-Anthropocene ontologies guide an exit of current approaches to diets? Considering temporality, what post-Anthropocene ontologies are possible in future diets for sustainable food systems? Through the ontological positions defining three distinct temporalities, considerations for guiding future diets in(to) the post-Anthropocene are proposed. Indigenous ontologies are presented as pre-Anthropocene examples that depict humans and non-humans in relational diets. Underlying Anthropocene ontologies define current unsustainable diets. These ontologies are described to present the context for the food systems challenges this article aims to address. A post-Anthropocene illustration then employs object-oriented ecosophy along with indigenous ontologies as theoretical foundations for shifting from the dominant neoliberal paradigm in current ontologies. Ontologically-based dietary guidelines for the post-Anthropocene diet present the ontological turns, consideration of temporality, and outline technological orientations proposed for sustainable future food systems. This is a novel attempt to integrate non-anthropocentric theories to suggest possible futures for human diets in order to exit the Anthropocene epoch. These non-anthropocentric ontologies demonstrate how temporal considerations and relational worldviews can be guidelines for transforming diets to address public health concerns, the environmental crisis, and socioeconomic challenges.

Keywords: sustainable diets; Anthropocene; indigenous ontologies; temporality; sustainable futures

---

1. Introduction

Climate change is challenging food systems, livelihoods, and human and ecosystem health [1]. There is high confidence that the global food system is the most predominant contributor to current environmental degradation [1]. Led by agriculture, planetary boundaries have been surpassed in biosphere integrity, biogeochemical flows, and land-system changes [2,3]. Globally, 40% of land is used for agriculture [4] with unprecedented rates of expansion intensifying productivity and supporting increased consumption [1]. Impacting marine ecosystems, fishing livelihoods, and sustainable fisheries, 60% and 30% of global fish stocks are completely exhausted or over-fished, respectively [5]. Pasture and cropland conversion have been chief causes of species extinction and deforestation [6], and cropland expansion results in larger declines in biodiversity [1]. Eutrophication from overuse of nitrogen and phosphorus [7] has been coupled with the consumption of 70% of the world's fresh water for agriculture [8]. Estimates of 25–30% of global greenhouse gas emissions (GHGs) are owed to livestock and agricultural production [9]. These GHGs have caused a rise in global temperatures, changes in precipitation, and a negative feedback loop impacting food systems [10].

Over the past several decades, the global food system has changed dramatically [11]. Increased demand for food, fuel, and fiber biomass "has been met by converting ecosystems into [global] production ecosystems" [12]. Highly-varied food production systems across the globe have shifted into supply chains that are increasingly more specialized, complex, and vertically integrated (i.e. corporations own intermediate means of production) [11]. With greater distances between producers and consumers, fewer people are growing their own food and more are buying from markets [11]. There is a transition from the direct consumption of raw ingredients to increased agricultural production for ultra-processed food ingredients [11]. Given such changes, the power of the private sector has increased, and labor, power, capital, and values have been concentrated in large agribusinesses and food industry [11]. At the same time, nutrition transitions in diets have set food production systems at odds with the provision of ecosystem services, increased the diet-related noncommunicable disease prevalence, and have contributed disproportionately to depletion of natural resources [13]. Reciprocally, many food system changes are driven or exacerbated by population growth, disparities in income distribution, urbanization, and dietary consumption practices [13].

The current environmental crisis has been much debated as a product of human action and has been called 'The Anthropocene' [14,15]. The Anthropocene is recognized as the current geological epoch catalyzed by substantial human impacts on the planet [16,17]. There is strong debate around the exact dates and definition of the Anthropocene concept [18–20]. Yet, the assertion of the Anthropocene as a distinct, human-centered geological era has pressed humanity to rethink our relationship to future generations and non-human entities in the world [19]. Turns in ontologies that underlie these relations are proposed as a novel means of approaching the end of the Anthropocene epoch [21].

Much literature focusing on sustainable future food systems give many, practical guidelines for how transformations can be made now. Such proposed actions include an increase in intensive production through efficient technological solutions, narrowing production yield gaps while minimizing negative externalities, avoiding overconsumption and food waste, and transforming diets to incorporate fewer animal-based, ultra-processed, and sugar-sweetened foods [13,17,22–25]. Though these are consistent and practical solutions, they fall short of addressing the deeper philosophical turn needed for humans to enact changes in reality.

Discussion of food systems and diets have generally only extended to 2050 [13,17,22,25]. To address the environmental crisis of the Anthropocene, several fundamental changes to the global food system and transformations in human action have been proposed. Such actions include full-supply chain policy interventions [26], redirection of finance for sustainability, radical transparency and traceability, and including keystone actors (e.g., transnational corporations) as global drivers of change [12]. However, movement beyond 2050 short-term recommendations will be needed for providing more temporally distant (i.e. several future generations) diets from future sustainable food systems.

The use of the term 'Anthropocene' is inherently an assertion of temporality and invokes the possibility of both a 'pre-' and 'post-Anthropocene'. Temporality means existing within and having relation to time. The timely issues of the Anthropocene related to sustainable diets have been addressed through the wisdom of indigenous peoples whose practices will exemplify a 'pre-Anthropocene' [27]. A time after the anthropocentric, consumptive dominance of the planet—a post-Anthropocene—is theorized as possible [28]. In this post-Anthropocene humans are de-centralized as the sole subjects of consideration in a sustainable food system. Such non-anthropocentric sustainable food systems remain within planetary boundaries where consumers have remade themselves more cognizant members of the global community through deeper ontological turns [11,17,29].

The transformation of current diets to sustainable diets has been widely promoted. Diet transformations have been proposed as one way to address the nutrition transition as well as the global food systems challenges contributing to the environmental crisis [13,17,22–24,30]. Diets are defined as the eating patterns across the lifespan and the types of foods consumed by a person habitually [31]. Sustainable diets are those with "low environmental impacts which contribute to food and nutrition security and to a healthy life for present and future generations" [32]. Since value-based social norms and self-efficacy will drive diet transformations more than pressures of perceived climate change or health risk [33], we argue deeper philosophical perspective shifts could help with the transition to sustainable diets.

Diets are one of the most profound, intimate connections humans have to their external environment. External realities—or ontologies—deal with the questions of existence and the nature of relations to what exists. The ontologies behind diets define the way people, as eaters, understand the nature of reality and their relation to that reality through the consumption of constitutive nutrients. Larger, systemic diet transformations may not come independently of turns in underlying world-views [34]. Possible approaches to the "require[d] radical shifts in deeply held values" [12] are suggested in this article through the guidelines for ontological turns (indigenous and object-oriented ecosophy), temporality considerations (past/future connection and present opportunity), and technological orientations (slow/low and high tech) for post-Anthropocene futures.

It must be recognized that the terms 'ontology', 'Anthropocene', and 'sustainability' are creations of the western academic canon. As Hunt [35] notes "western ontological possibilities are bounded in ways that limit their ability to fully account for indigenous worldviews". Recognizing these limitations, the terms are used here in the western context as a means to work towards commensurate discussions of the current environmental crisis—widely agreed upon by western scientists [36]. Through learning from indigeneity, western scholarship can go beyond current ontological limits for turning from anthropocentric worldviews [35].

Indigenous ontologies are proposed to respond to anthropocentric challenges we face. Indigenous ontologies outline diets where foods have significant relationships with human and nonhuman communities in temporally deep, spatially local, and complex ways [37]. Such ontological outlines are presented to give current consumption practices the ability to move to temporally distant consideration needed to exit the Anthropocene [21].

To connect indigenous ontologies to western academic contexts that have already proposed an Anthropocene exit, object-oriented ecosophy is also used in this article. Developed from object-oriented ontology and ecological philosophy, object-oriented ecosophy parallels much of pre-Anthropocene indigenous ontologies [21]. The theory of object-oriented ontology radically asserts that worldviews cannot solely consider human subjects but must also encompass other objects (e.g., crops, crude oil, oceans) and their fundamental characteristics [38,39]. Object-oriented ecosophy presents an outline for the transition to ontologies which both de-center humans and consider the relationality of objects [21].

There is a current—spatial and temporal—disconnect among the people, places, and things consumed and those entities impacted by that consumption [40]. The food system of the Anthropocene is predicated on anthropocentrism, excessive consumption, negative human and non-human externalities, and an irrational separation of actions and consequences [10,40,41]. The era of post-Anthropocene will need non-anthropocentric philosophies, practices, and institutions. We will need sustainable relations among humans and non-humans and internalization of the effects of anthropogenic influences. To exit this Anthropocene era, a deliberate connection of the time and space of current and future diets—including the food systems which provide those—will need interdisciplinary understanding and approaches that go beyond any one paradigm or system [20].

Assuming we are in the Anthropocene epoch and that we must exit this era for a sustainable future, the purpose of this article is to argue that turns in ontologies are needed, informing and driving transformations in diets. We conceptually address two central research questions: how can pre-Anthropocene ontologies guide an exit of current approaches to diets? And, considering temporality, what post-Anthropocene ontologies are possible in future diets for sustainable food systems? To advance conceptual discussions of future diets, this paper draws upon literature describing indigenous food systems and the unsustainable Anthropocene context. We advance this discussion to address ontological turns for the post-Anthropocene. The theories and examples of indigenous worldviews paralleled with object-oriented ecosophy are used herein. This article addresses the philosophical paradigm shift needed to exit the Anthropocene through the conceptual discussion of eating, thinking, and being constituted of The Post-Anthropocene Diet.

2. The Pre-Anthropocene Diet: Indigenous Ontologies

2.1. *What Is Pre-Anthropocene?*

Sustainable food systems have already been in practice for millennia [27]. Indigenous food systems offer answers to the questions scientists and policymakers are asking today. How do we generate food for people while also maintaining natural resources, the environment, and biodiversity? How do we efficiently and sustainably use energy (i.e. food as caloric value) within the system? What multipurpose strategies build capacity for the generation of byproducts, shelter, and medicines?

The pre-Anthropocene is exemplified here through pre-colonial, indigenous ways of being and knowing [35]. We will call these ways of being and knowing indigenous ontologies. Indigenous ontologies underpin the diets of indigenous peoples, which are linked to food systems that provision those diets [37,42]. From the exploration of local foods to the incorporation of sustainable commercial foods, indigenous ontologies can model sustainable diets for larger populations [27,43]. Indigenous food systems are a globally-varied, diverse set of indigenous peoples' management, traditional practices, and temporally-deep cultural knowledge [27]. Such food systems generate food from the respective territories of indigenous communities.

The current environmental crisis is a challenge for all of humanity but a historically predominant challenge for indigenous communities. Whyte [37] claims that "in the Anthropocene, some indigenous peoples already inhabit what our ancestors would have likely characterized as a dystopian future". Whyte argues that the environmental crisis and destabilization is only the most recent challenge facing indigenous peoples, with their Anthropocene ignited by colonialism. Climate change is an issue of contemporary society. Yet, the environmental crisis is a "historically brief, highly disruptive moment" of the many and longstanding anthropogenic threats to indigenous peoples [37]. Such anthropogenic threats have served to systematically and rapidly force indigenous peoples to adapt or lose relationships with plants, animals, and ecosystems of their ancestors [37]. What western modernity characterizes as future dystopias of climate disaster—critically threatened species and ecosystems—is the present-day dystopia of indigenous ancestors [37]. Exiting the Anthropocene will require shared and saved ontologies for present action and future generations. We posit post-Anthropocene ontologies should parallel indigenous pre-Anthropocene.

2.2. *What Are Pre-Anthropocene Ontologies?*

Indigenous ontologies—though globally distinct and diverse—define indigenous diets. Provisioning diets is one of the primary interactions with non-human beings that comprise indigenous food systems. Indigenous ontologies form food systems through a holistic view that de-centers human beings and the production of food itself [27]. An indigenous food system encompasses "all food within a particular culture available from local natural resources ... . It also includes the sociocultural meanings, acquisition/processing techniques, use, composition and nutritional consequences for the people using the food" [44]. A further distinguishing feature of indigenous food systems is that they instinctively avoid commercial orientation, combining shared production and consumption [45].

The ontological view of not differentiating between the needs of people and the environment alters the relationality inherent in indigenous diets. "We do not have the right to interfere with water's duties to the rest of Creation," asserts Anishinaabe scholar Deb McGregor [46]. In this First Nations perspective on environmental justice, McGregor outlines an ontology where humans are not centered as the sole beings of ethical consideration. In the verb-based relational languages framing indigenous ontologies, the subject of the diet (i.e. food item) is made into an object alongside the eater (e.g., an apple tree is 'tree-ing', as a person would be 'be-ing') [47].

Food systems are complex, and indigenous ontologies recognize the relationality and connectedness of individuals, technology, and society. In the pre-Anthropocene, technological orientation differed from current practices through the use of slower and lower-tech means of ecological management for food provisioning [27]. Food preferences were accounted for given the deeply relational aspects of social and cultural structures [27]. Societies evolved around ancestral land stewardship and acknowledgment of the necessity of the care and connection to all other human and non-human beings [37].

The dominant colonial paradigm of ownership, borders, and superiority is challenged and dismantled by indigenous food systems [27]. What commercial food systems—and most human (extr)activities—regard as natural resources, indigenous ontologies perceive as spiritually embedded and intrinsic to sustaining relations with living and non-living beings [27]. Listening to and learning from the indigenous ontologies informed by the hundreds of years of their Anthropocene is valuable in that these ontologies "are not based on the dread of certain futures. Rather, they arise from indigenous perspectives on how to respond to anthropogenic climate destabilization based on having already lived through local losses of species and ecosystems" [37].

2.3. *How Do Pre-Anthropocene Ontologies Consider Temporality?*

Indigenous ontologies are temporal in practice. Such a practice defines relational consideration of temporally distant and proximal human and non-human beings. Relations, agency, and actions are considered through the 'seventh generation' pre-ceding and pro-ceding the present community [48]. The indigenous conception understands time as a relational opportunity in which to act [44]. Choices made in the present create a future. The future will someday be a part of ancestral history. Indigenous food systems, and diets by extension, already encompass this type of eating. The ancestrally relational quality of foods is the focus. There is the recognition that each being (human and non-human) is part of a temporally-deep system [27]. New conceptions of consumption through a theory of time and relationality can transform present diets to exit the Anthropocene. Through ancestrally-deep indigenous ontologies, we may be able to prevent consuming our planet's futures today [49].

2.4. *What Are Examples of Pre-Anthropocene Diets?*

Relational indigenous ontologies can be seen in the diets of pastoralists and nomadic hunters and fishers [27]. Central African and Asian hunters and gatherers have a code deeply rooted in the traditional knowledge of the community. Such codes form a food system composed of reticular spaces and nodular relations. Nodes are the collective points connecting humans and non-humans in the food system (e.g., spiritual and cultural community events, ceremonies, and conversations over feasts). Reticular spaces are places with different functions which cannot be understood without consideration of relationality (e.g., sacred fruit harvesting areas, emergency feeding spaces, tuber provision for/with neighboring communities) [27]. The food systems are maintained through relational mechanisms. Stratified, complex exchanges among elders, adults, and children delineate different tasks and knowledge for each group and each reticular area [27].

The ontological practice of indigenous diets promotes the preservation of ecosystems. For example, indigenous-led public events feature restoration programs of sturgeon, wild rice, and water. These public events have brought together members of settler society with indigenous people to learn about the importance and relationality of humans to the rest of their environment [37]. Diets incorporate territory-food linkages, cultural and spiritual relations, and traditional knowledge [27]. Such indigenous practices have been proposed as more efficient and sustainable than present agricultural methods [27].

3. The Anthropocene Diet: Anthropocentric Ontologies

3.1. *What Is the Anthropocene?*

The Anthropocene geological epoch is the era of human-driven impacts on the Earth System. There is a conflicting discussion on when exactly the Anthropocene officially started: in the Industrial Revolution with the steam engine [18], since early-twentieth-century global temperature rise [50], or in 1950 with the age of the atomic bomb post-WWII [16]. Despite how the beginning is defined, Anthropocene discourse consistently centers humans as the agents of geological change [15,16,51]. Human impacts have significantly altered global biogeochemical flows, atmospheric conditions, ecosystems, landscapes and oceans [2,14,15]. The human-centered era has influenced the global climate through activities that increase the levels of carbon dioxide and greenhouse gasses in the atmosphere, melt ice sheets, raise sea levels, and severely impact global biodiversity [15,16,51].

The Anthropocene describes an unsustainable structure and evolution of the food system. There are many main challenges, which describe the food system of the Anthropocene [17]. Food production is the largest anthropogenic pressure on Earth, causing threats to local ecosystems and global Earth System stability [17]. Risks to people and the planet are exacerbated by population growth and current trends in diets [17]. Though the concept and definition of the Anthropocene are debated, the anthropocentric nature and human-driven impacts of the current era are beyond certainty [36]. There is high confidence that the Earth System is past the point of any return to pre-industrial conditions [1].

3.2. *What Are Anthropocene Ontologies?*

The ontologies underpinning the Anthropocene are foundationally human-centric. The discourse around the Anthropocene asserts humans as the cause and agents of change [14,18,20]. Humans are the central agents and central beings impacted, which defines anthropocentrism. Anthropocentrism reinforces human-nature dualism and distinguishes humans as separate from nature.

As opposed to such 'Anthropocene ontologies', ecocentric theories recognize all human actions and values as situated within and subordinate to the global ecosystem [52,53]. We acknowledge that human-centered worldviews may lead to similar sustainability transitions. However, we propose an ecocentric approach as a first step in broadening worldviews. Ecocentrism de-centralizes humans in recognition of the central importance of the non-human world.

The human-nature, subject-object dualism in ecocentric thought is often problematized. Such privileging of humans as subjects over nature objects perpetuates the human-nature distinction. A distinction which is cited as a defining issue in the Anthropocene [21,37]. What such ecocentric perspectives lack is the consideration of equality and relationality of humans and non-humans.

We propose two ontological approaches here in response to the human-centered, neoliberal Anthropocene ontologies. Indigenous ontologies remove humans from their dominant, anthropocentric agency [54]. The indigenous approach reflects equality through recognizing that water, foods, ecosystems, etc. are living entities with rights to live and rights to not have their duties to other beings interfered upon by human action [37,46]. Similarly, object-oriented ecosophy "avoid(s) the human-nature dualism by considering each thing an object while still arriving at an ecologically relevant view of reality" [21].

3.3. *How Do Anthropocene Ontologies Consider Temporality?*

The consideration of temporality in Anthropocene ontologies is conspicuously short-sighted. The effects on the global environment have escalated in the past three centuries due to the rapid expansion of humankind, both in population and the gratuitous exploitation of natural resources per capita [14]. Decisions made in the Anthropocene have been characterized by a disregard for possible future impacts. The disregard for future temporalities in the Anthropocene has privileged uninhibited growth. A growth that has out-paced ecological boundaries for human and non-human wellbeing and equity on the planet [2,15,55]. Ontological turns toward decisions made with future temporal realities in mind may catalyze the end of the Anthropocene. Such considerations of the future define sustainable, healthier, climate-secure, and ethical diets.

3.4. What Are Examples of Anthropocene Diets?

Diets in the Anthropocene are exemplified by the current, unhealthy, unsustainable nutrition transition. With the advent of colonialism, industrialization, globalization, and heavy processing, diets have transitioned [37,56,57]. Nutrition transitions have been seen across the globe to diets high in calories, heavily-processed, and animal-based foods with deficiencies in balance, diversity, and adequacy [11,17,57]. Without transformations away from the current dietary practices, there will be further increases in diet-related non-communicable diseases (e.g. obesity, heart disease, diabetes) and irreversible environmental degradation [17,23,24,58]. In transforming diets out of the Anthropocene, we need a paradigm shift to considering foods as objects in themselves. We propose placing decisions around food consumption and production in temporalities that seek post-Anthropocene realities.

4. The Post-Anthropocene Diet: Ecosophical Ontologies

4.1. What Post-Anthropocene Is Possible?

To exit the Anthropocene, we propose turns from the current anthropocentric ontologies. Though difficult to achieve, ontological turns will have to be made to exit the dystopia that is the Anthropocene. Ontologies underlying diets would outline consumption in the present while also balancing the consideration of the future. The dietary guidelines proposed here are not about food groups and portions. We propose guiding considerations that enable the epochal changes needed to exit the Anthropocene ontologically, temporally, and technologically.

Complexities will be present in post-Anthropocene food systems, which future diets will need to address. Context-specific environmental and socio-economic factors will be relevant to the post-Anthropocene diet [59]. Local food cultures, production possibilities, and seasonality must drive sustainable diets [59]. Supplementation and imported products will need to be coupled with potential novel food technologies to consider global production efficiencies and nutrient sufficiency [59,60]. Food security will remain an important consideration for sustainable post-Anthropocene food systems and diets [61,62]. Any socio-cultural, technical, or ontological transformations require consideration of the potential unintended consequences through interdisciplinary and multi-sectoral collaborations [59].

4.2. What Are Possible Post-Anthropocene Ontologies?

Ending the Anthropocene will mean turns in ontologies such as to object-oriented ecosophies. Through extending the theories of object-oriented [38,39] and ecological philosophy [63,64], Heikkurinen et al. [21] outline an 'object-oriented ecosophy'. Such an approach is used here to suggest ontological outlines for turning conceptions and consumption to post-Anthropocene diets.

Object-oriented ecosophy—mirrored previously in indigenous ontologies—illustrates three essential qualities of objects. Foods, and the systems that produce them, are seen as autonomous, intrinsic, and unique [21]. These essential qualities have both theoretical and practical implications for diet transformations central to the post-Anthropocene diet. All objects (i.e. foods) have a degree of autonomy, some more than others. This autonomy may be used to assign or explain varying degrees of moral agency [21]. The quality of intrinsicality implies that no object should be treated as a means, but they are ends in themselves. Recognizing intrinsicality releases objects from instrumental rationale or use without deeper value [21]. The uniqueness of objects recognizes their irreducibility and non-substitutability. This quality suggests that every object occupies a specific place and time, which is important for embracing the diversity of objects for organization and conservation [21].

What the application of object-oriented ecosophy means for diets is that foods and the environments that produce them are considered as objects—with relations inherent to other objects. The post-Anthropocene ontological turn of object-oriented ecosophy would erase the divisions of humans and non-humans. This ontological foundation proposes foods not merely as objects of consumption. Foods are part of the world-in-formation as they are intertwined in the process of becoming [65]. Such turns in ontology around consumption are proposed here to de-centralize humans. This would be a paradigm shift in diets for sustainable future food systems.

4.3. How Would Post-Anthropocene Ontologies Consider Temporality?

New questions of temporality will also define post-Anthropocene ontologies and diets. The application of indigenous and object-oriented ecosophy provides non-anthropocentric temporal outlines for exiting the Anthropocene. What this means practically is that, through our choices, there is an imagined narrative of the future that is interpreted and defined by our present reality [49].

To give an example of temporality considerations in turning ontologies, Robinson [49] presents the narrative of driving a car down a street. There appears a child playing next to the street, which alters the present reality when the driver conceives of a possible future where the child runs onto the road. The present is now redefined as a dangerous situation, requiring corrective, anticipatory actions where the driver slows down. Inserting existential threat, the driver temporally re-conceptualizes the very decision of purchasing the car in the first place to avoid the possibility of ever endangering that child [49]. In this scenario, present eaters and their diets are the drivers, and the future planet, people and all other non-human beings are the endangered child.

To give an example of the temporal consideration of post-Anthropocene thinking in diets we present a possible re-interpretation of consuming high GHG emissions-producing foods. Purchasing and consuming GHG emissions-intensive foods is given new meaning by the environmental crisis. The eater recognizes a possible future where climate change has devastated agriculture and compromised future generations' ability to grow food. This implies that future turns in ontologies have the ability to alter how we interpret "instrumental anticipatory consumption acts in the present" [49]. A post-Anthropocene diet would situate the eater in a chronologically responsible present, determined by the past and determining the future.

4.4. What Would Be Examples of Post-Anthropocene Diets?

Assuming humans will still be present in a post-Anthropocene, what would diets and their considerations look like? A post-Anthropocene diet would act and select foods that throw off the increasingly dominant, destructive capitalist tendencies of the Anthropocene. In practice, this post-Anthropocene diet perspective would be a re-definition of consumption. New definitions would incorporate and move toward activities that contribute to broader human and non-human outcomes. Choices would re-orient the eater to considering and deepening the meaning and practice of consumption. This means the selection of foods with other social benefit outcomes or from technological innovations with more efficient means of production. Technological solutions already exist for sustainable food options (e.g., vertical farming, cell-cultured meat, plant-based protein alternatives). More technologies will be developed which will allow for many different possible future diets from sustainable food systems.

A second example of a possible post-Anthropocene diet takes a slightly different approach, reconsidering the act of consumption altogether. Changes will not be led by technological innovations, but slower, lower-tech solutions will arise/reemerge. There will be a concentration on not only less consumption but consumption that arranges living frugally. Such concentration would adjust societal metabolism to centrally focus on foods as objects with their own essential qualities. The food systems provisioning diets will transform to de-centralize humans for a world where production and consumption are downscaled. Sufficient, slower food options of hunting and gathering will supplant the faster foods of efficiency. Consumers will step away from the globalized markets of superstores and processed foods. Foods may be canned and processed once gathered, grown oneself, or shared with others in times of plenty. These actions would reconsider the temporality of diets.

5. Ontologically-Based Dietary Guidelines

A summary of the three temporal epochs presented through their ontologies, consideration of temporality in decision making, and practical examples of diets (Sections 2–4) is outlined in Table 1. The pre-Anthropocene is presented through indigenous worldviews. The Anthropocene defines a temporal worldview that disregards future generations through unsustainable diets. A proposed post-Anthropocene ontological turn is presented as paralleling the indigenous pre-Anthropocene. Such turns can work to define a new era through object-oriented ecosophy, reconsideration of temporality, and practical or technological transformations in diets.

Table 1 presents possible outlines for turns in ontologies to theorize and realize sustainable futures. This is partly an application of what Heikkurinen et al. [21] offer as an ontological outline—their object-oriented ecosophy—for the transition to a post-Anthropocene society. For them, this means the transition to ecological organization theory and the practice that reimagines object-object relations for "the peaceful coexistence of objects" [21].

For diets, this means using the inherent qualities of objects to reduce the instrumentalization of foods and natural resources. These diets outline an ontological future which releases the eater from anthropocentrism. It is proposed here that such an ontological turn is needed to reach a sustainable post-Anthropocene.

We recognize humans will still need to eat and use resources for the provisioning of those foods. Post-Anthropocene diets are proposed as those which reduce the bias of yield maximization, agricultural industrialization, and commercial food production. We propose an indigenous or object-oriented ecosophy that will position consumers with ontologies to catalyze a philosophical paradigm shift. Non-anthropocentric, pre-Anthropocene diets have been exemplified in indigenous food systems. These diets are informed by an indigenous ontology that has an inherent and relational understanding of how local foods are adapted to local environments.

It has been asserted that the indigenous ways of eating are more resilient in the face of climate challenges [27]. As ontologies turn, post-Anthropocene diets would be composed of fewer industrially monocultured foods. The post-Anthropocene diet will transition from foods furnished through unsustainable agricultural practices of the global-industrial Anthropocene. Diets will consist of sustainably produced, gathered, hunted, or fished foods. Consumption patterns would be led by seasonality and availability, which, though obvious, could be drastically different in a post-Anthropocene world given changing global climate conditions.

As a presentation of possible ways forward, we posit 'Ontologically-based dietary guidelines.' Potential ontological turns, temporality considerations, and technological orientations are recommended as guidelines for the post-Anthropocene diet. Firstly, the ontological turn requires a de-centering of humans. Such de-centering necessitates an understanding of the relationality of foods as objects given their essential qualities: autonomy, intrinsicality, uniqueness and/or through indigenous worldviews [21,44]. Secondly, diets may be considered through consideration of temporality: the questions of when and how much to act or consume are raised. The temporal considerations that embody the post-Anthropocene diet will be guided by asking (i) for what quantity of time in the future is this decision made for a sustainable future food system—to exit the Anthropocene? and (ii) for what quality of present or future is this decision to consume?

Table 1. Pre-, post-, and Anthropocene epochal descriptions.

Epoch	Pre-Anthropocene	Anthropocene	Post-Anthropocene
Ontologies/Worldviews	Indigenous wisdom (histories, stories, languages, artistry, spirituality), relational, connected to the land and all beings, all beings have rights in and of themselves, humans should not interfere with the duties of other beings to each other, protecting natural resources	Human-centered, perpetuated human-nature dualism, agential anthropocentrism, neoliberal, colonial, productivist, efficiency	Object-oriented ecosophy, systemic, complex and adaptive, all objects have intrinsicality, autonomy, and uniqueness, equitable ecocentrism, de-/post-colonial, sufficiency
Consideration of Temporality	Connected to ancestors and to progenitors, decisions are made for the seventh generation	Shallow, immediate gratification, efficient, future generations supplanted for present consumption	Present consumption regards the needs of/possible impacts on the future, is cognizant of historical context
Examples of Diets in practice	Wild food, hunting, gathering, food preservation (drying, salting, smoking, etc.), pastoralists, nomadic, identification, soil maintenance, ancestral seeds, medicinals, cultural cooking techniques	Highly processed, energy-dense, nutrient-deplete, sugar-sweetened, globalized, heavy carbonization, contributing to/leading the environmental crisis, cheap, fast foods, convenience, high food wastage	Local, seasonal, foraging, reduce food waste, 'sustainable', plant-based, within planetary boundaries, growing own food/permaculture, slow food, return to traditional/culturally appropriate, affordable, soil regenerative, technologically-produced ecological diets

We present low- and high-tech examples for understanding technological orientation in future temporalities. A slow or low-tech future may embrace gardening, canning, drying, and preserving foods. Sustainable production and extensive farming systems may be combined with hunting, gathering, and foraging for wild foods. Localization of markets and community-supported agriculture are models that embrace slower food futures. The technological orientation of high-tech futures may embrace sustainable production models of cellular agriculture, vertical farming, biotechnology, and 'smart' agro-technology. In-home or large-scale biodigesters that run on renewable energy and produce biofuels may become more common. These examples are presented around a future diet situated in high-tech and slow/low-tech solutions. Neither example is necessarily superior to the other but would both be equally possible outcomes of this line of reasoning and ontological turn for the post-Anthropocene diet. Future sustainable food systems would most likely consist of some amalgamation of both low- and high-tech futures.

There are limitations to adopting these worldviews and making decisions that redefine food systems outside of efficiency and productivity. It takes time and money to establish technologies for sustainable consumption. Further, there is not one single, simple, or all-encompassing technology that will transform diets for health and sustainability. Using cell-cultured meat as an example, several negative environmental outcomes can be reduced, but this technology requires large amounts of energy and comes at a high cost [66,67]. There are also barriers to entry and access to such technologies. Lack of individual knowledge and funding for more research are limitations in this field. Moreover, dependency on technological solutions still ties consumers to markets or spaces of production. This dependency may be lessened through owning smaller, home-based means of technological production, but again these come with barriers of money and resources.

It is also recognized that there are many challenges and restrictions on individual decision making for transforming diets. Many people are restricted by time, space, and knowledge to gather, grow, cook, and process their own foods. However, this article is set to challenge the centrality of constant growth, efficiency, and productivity and embraces now as the time to make transformations even if small but sustainable. Current diets can take small steps to de-centering humans in the food system. We can choose to move towards this epochal exit through consumption. Choices that leave more space for temporally and spatially distant humans and non-humans in the world to be their autonomous, intrinsic, unique selves.

The feasibility of ontological turns compared to more engineered steering of the global food system must also be considered. Policy interventions are possible. Yet, the correct actors need to be targeted, knowledge of effective changes is limited, and enforcement can be a challenge [26]. Redirecting finance and engaging keystone transnational corporations can be key leverage points for systemic transformations in global production ecosystems [12]. However, there is often an opaque link between financial flows and environmental change [12]. Such neoliberal expansion led by large corporations is asserted as a main driver of environmental degradation of the Anthropocene [10,40,41]. Radical transparency and traceability in sustainability issues may be influential in aligning consumer purchasing with sustainable thinking [12]. However, given "the urgency and complexity of this challenge" [12], the transformative change will also require radical shifts in current economic paradigms. Such change cannot come independently of transformations in deeply held values, education, and social structures underpinning consumer behaviors [12]. Ontological turns de-centering humans in diets presents one option for the feasible unfolding of reality where consumption behaviors can influence health and sustainability outcomes [21,68].

6. Conclusions

We conceptually addressed two central research questions: how can pre-Anthropocene ontologies guide an exit of current approaches to diets? Considering temporality, what post-Anthropocene ontologies are possible in future diets for sustainable food systems? This paper asks eaters to question, 'How should we make dietary choices (e.g., eat/consume) in the present and also balance consideration of the future in a way that works to exit the Anthropocene?' As an answer, we present ontologically-based guidelines for the post-Anthropocene diet, proposing a philosophical paradigm shift needed to exit the Anthropocene. Through the conceptual discussion of eating, thinking, and being, a suggestion of how the Anthropocene might come to an end is made. Indigenous ontologies and object-oriented ecosophy are invoked to turn ontologies.

There are broader implications of this conceptual model for future diets and research. This article contributes an ontological perspective to the growing discussion (and debate) around the Anthropocene. Discussion of indigenous ontologies and object-oriented ecosophies adds novel contributions to conceptual papers of future sustainable food systems. Recent conversations around sustainable dietary recommendations have largely disregarded philosophical transitions. We hope to help initiate deeper considerations. We also add to a mostly natural sciences-based discussion of the Anthropocene through the interdisciplinary, conceptual approach of this work. There is a recognized need for both the natural and social sciences in facing the challenges of modernity and moving to sustainable futures.

Limitations of this approach include consideration of only temporality. Temporality as the outline for this discussion prevented the full consideration of the relational and spatial aspects of diets and food consumption. More discussion of relations within and among humans and food objects is needed. Complex systems theories may add to discussions of food systems and relationality. Spatial considerations of how food is grown and distributed in the globalized economy would add to this discussion. Deeper understanding, informed by more dimensions of reality, will allow for further turns of ontologies. More work is needed to find ways to practically apply theorizing presented here to go beyond philosophical navel-gazing.

We also recognize the potential of idealizing or appropriating indigenous ontologies. The intention here was to present indigenous ontologies, not as one, all-encompassing, distinct worldview. We recognize the myriad indigenous ways of knowing and seeing the world. These indigenous ontologies are used as edifying examples that work in the world and have practice consistently dealing with anthropocentric climate and cultural destabilization. The many indigenous worldviews are not to be romanticized or exoticized. Often neglected indigenous worldviews should be seen as dynamic contributions to the global discussion of how to live in and face the challenges of the present dystopic environmental crisis.

We present one set of guiding considerations to enable the epochal transformation needed to exit the Anthropocene. This work gives practical considerations for turning ontologies with examples of indigenous ontologies and object-oriented ecosophy applied to diets. In reality, such ontological turns are not so simple and practical. This work suggests a small piece for the larger puzzle of moving toward sustainable futures.

Suggesting we change worldviews is strikingly easier than actually changing them. Changes to education, policies, economic and social structures are required. More research on how to exit this epoch and how to turn ontologies is needed. This article suggests guidelines for one place to start. The conceptualization of the post-Anthropocene diet in this article is just one presentation of a small 'slice' of the larger model of ontologically-based dietary guidelines. There is much future work to be done to move eaters to diets for future sustainable food systems.

Author Contributions: Conceptualization, R.M.; investigation, R.M.; writing—original draft preparation, R.M.; supervision, H.L.T.; writing—review and editing, R.M., H.L.T.; visualization, R.M., H.L.T. All authors have read and agreed to the published version of the manuscript.

Funding: This research was funded by the Research Funds of the University of Helsinki.

Acknowledgments: We want to acknowledge that this work was supported by discussions and encouragement from Pasi Heikkurinen out of the Department of Economics and Management, University of Helsinki, Helsinki, Finland and the feedback of group members in the Culture and the Crisis seminar series including Markus Vinnari and Toni Ruuska.

Conflicts of Interest: The authors declare no conflict of interest.

References

1. Intergovernmental Panel on Climate Change. *Climate Change and Land: An IPCC Special Report on Climate Change, Desertification, Land Degradation, Sustainable Land Management, Food Security, and Greenhouse Gas Fluxes in Terrestrial Ecosystems*; Intergovernmental Panel on Climate Change: Geneva, Switzerland, 2019.
2. Steffen, W.; Richardson, K.; Rockström, J.; Cornell, S.E.; Fetzer, I.; Bennett, E.M.; Biggs, R.; Carpenter, S.R.; De Vries, W.; De Wit, C.A. Planetary boundaries: Guiding human development on a changing planet. *Science* **2015**, *347*, 1259855. [CrossRef]
3. Campbell, B.M.; Beare, D.J.; Bennett, E.M.; Hall-Spencer, J.M.; Ingram, J.S.; Jaramillo, F.; Ortiz, R.; Ramankutty, N.; Sayer, J.A.; Shindell, D. Agriculture production as a major driver of the earth system exceeding planetary boundaries. *Ecol. Soc.* **2017**, *22*, 8. [CrossRef]
4. UN FAO. *FAOSTAT Land Use*; UN FAO: Rome, Italy, 2019.
5. UN FAO. *The State of World Fisheries and Aquaculture*; UN FAO: Rome, Italy, 2018.
6. Tilman, D.; Clark, M.; Williams, D.R.; Kimmel, K.; Polasky, S.; Packer, C. Future threats to biodiversity and pathways to their prevention. *Nature* **2017**, *546*, 73–81. [CrossRef] [PubMed]
7. Diaz, R.J.; Rosenberg, R. Spreading dead zones and consequences for marine ecosystems. *Science* **2008**, *321*, 926–929. [CrossRef] [PubMed]
8. Molden, D. *Water for Food Water for Life: A Comprehensive Assessment of Water Management in Agriculture*; Routledge: Abingdon, UK, 2013.
9. Tubiello, F.N.; Salvatore, M.; Cóndor Golec, R.D.; Ferrara, A.; Rossi, S.; Biancalani, R.; Federici, S.; Jacobs, H.; Flammini, A. *Agriculture, Forestry and Other Land use Emissions by Sources and Removals by Sinks*; Statistics Division, Food and Agriculture Organization: Rome, Italy, 2014.
10. Vermeulen, S.J.; Campbell, B.M.; Ingram, J.S. Climate change and food systems. *Annu. Rev. Environ. Resour.* **2012**, *37*, 195–222. [CrossRef]
11. Haddad, L.; Hawkes, C.; Waage, J.; Webb, P.; Godfray, C.; Toulmin, C. *Global Panel on Agriculture and Food Systems for Nutrition: Food Systems and Diets: Facing the Challenges of the 21st Century*; Springer: Berlin/Heidelberg, Germany, 2016.
12. Nyström, M.; Jouffray, J.; Norström, A.V.; Crona, B.; Jørgensen, P.S.; Carpenter, S.R.; Bodin, Ö.; Galaz, V.; Folke, C. Anatomy and resilience of the global production ecosystem. *Nature* **2019**, *575*, 98–108. [CrossRef]
13. Godfray, H.C.J.; Crute, I.R.; Haddad, L.; Lawrence, D.; Muir, J.F.; Nisbett, N.; Pretty, J.; Robinson, S.; Toulmin, C.; Whiteley, R. The future of the global food system. *Philos. Trans. R. Soc. Lond. B Biol. Sci.* **2010**, *365*, 2769–2777. [CrossRef]
14. Crutzen, P.J. Geology of mankind. In *A Pioneer on Atmospheric Chemistry and Climate Change in the Anthropocene*; Crutzen, P.J., Ed.; Springer: Berlin/Heidelberg, Germany, 2002; pp. 211–215.
15. Crutzen, P.J. The "anthropocene". In *Earth System Science in the Anthropocene*; Crutzen, P.J., Ed.; Springer: Berlin/Heidelberg, Germany, 2006; pp. 13–18.
16. Williams, M.; Zalasiewicz, J.; Waters, C. The Anthropocene: A geological perspective. In *Sustainability and Peaceful Coexistence for the Anthropocene*; Heikkurinen, P., Ed.; Routledge: Abingdon, UK, 2017; pp. 16–30.
17. Willett, W.; Rockström, J.; Loken, B.; Springmann, M.; Lang, T.; Vermeulen, S.; Garnett, T.; Tilman, D.; DeClerck, F.; Wood, A. Food in the anthropocene: The EAT–lancet commission on healthy diets from sustainable food systems. *Lancet* **2019**, *393*, 447–492. [CrossRef]
18. Zalasiewicz, J.; Williams, M.; Smith, A.; Barry, T.L.; Coe, A.L.; Bown, P.R.; Brenchley, P.; Cantrill, D.; Gale, A.; Gibbard, P. Are we now living in the anthropocene? *Gsa Today* **2008**, *18*, 4. [CrossRef]
19. Lewis, S.L.; Maslin, M.A. Defining the anthropocene. *Nature* **2015**, *519*, 171–180. [CrossRef]
20. Heikkurinen, P.; Ruuska, T.; Wilén, K.; Ulvila, M. The anthropocene exit: Reconciling discursive tensions on the new geological epoch. *Ecol. Econ.* **2019**, *164*, 106369. [CrossRef]
21. Heikkurinen, P.; Rinkinen, J.; Järvensivu, T.; Wilén, K.; Ruuska, T. Organising in the anthropocene: An ontological outline for ecocentric theorising. *J. Clean. Prod.* **2016**, *113*, 705–714. [CrossRef]
22. Foley, J.A.; Ramankutty, N.; Brauman, K.A.; Cassidy, E.S.; Gerber, J.S.; Johnston, M.; Mueller, N.D.; O'Connell, C.; Ray, D.K.; West, P.C. Solutions for a cultivated planet. *Nature* **2011**, *478*, 337–342. [CrossRef] [PubMed]
23. Tilman, D.; Clark, M. Global diets link environmental sustainability and human health. *Nature* **2014**, *515*, 518–522. [CrossRef]

24. Springmann, M.; Godfray, H.C.J.; Rayner, M.; Scarborough, P. Analysis and valuation of the health and climate change cobenefits of dietary change. *Proc. Natl. Acad. Sci. USA* **2016**, *113*, 4146–4151. [CrossRef]
25. Kummu, M.; Fader, M.; Gerten, D.; Guillaume, J.H.; Jalava, M.; Jägermeyr, J.; Pfister, S.; Porkka, M.; Siebert, S.; Varis, O. Bringing it all together: Linking measures to secure nations' food supply. *Curr. Opin. Environ. Sustain.* **2017**, *29*, 98–117. [CrossRef]
26. Kanter, D.R.; Bartolini, F.; Kugelberg, S.; Leip, A.; Oenema, O.; Uwizeye, A. Nitrogen pollution policy beyond the farm. *Nat. Food* **2019**, *2019*, 1–6. [CrossRef]
27. Kuhnlein, H.; Eme, P.; de Larrinoa, Y.F. 7 indigenous food systems: Contributions to sustainable food systems and sustainable diets. In *Sustainable Diets: Linking Nutrition and Food Systems*; CAB International: Boston, MA, USA, 2018; Volume 64.
28. Sexton, A.E. Eating for the post-Anthropocene: Alternative proteins and the biopolitics of edibility. *Trans. Inst. Br. Geogr.* **2018**, *43*, 586–600. [CrossRef]
29. Mansfield, B. Gendered biopolitics of public health: Regulation and discipline in seafood consumption advisories. *Environ. Plan. D Soc. Space* **2012**, *30*, 588–602. [CrossRef]
30. Aleksandrowicz, L.; Green, R.; Joy, E.J.; Smith, P.; Haines, A. The impacts of dietary change on greenhouse gas emissions, land use, water use, and health: A systematic review. *PLoS ONE* **2016**, *11*, e0165797. [CrossRef]
31. U.S. Department of Health and Human Services; U.S. Department of Agriculture. *2015–2020 Dietary Guidelines for Americans*; Dietary Guidelines: Washington, DC, USA, 2015.
32. UN FAO. *International Scientific Symposium: Biodiversity and Sustainable Diets United Against Hunger: 3–5 November*; UN FAO: Rome, Italy, 2010.
33. Eker, S.; Reese, G.; Obersteiner, M. Modelling the drivers of a widespread shift to sustainable diets. *Nat. Sustain.* **2019**, *2*, 725–735. [CrossRef]
34. Lang, T. Food control or food democracy? Re-engaging nutrition with society and the environment. *Public Health Nutr.* **2005**, *8*, 730–737. [CrossRef] [PubMed]
35. Hunt, S. Ontologies of indigeneity: The politics of embodying a concept. *Cult. Geogr.* **2014**, *21*, 27–32. [CrossRef]
36. Ripple, W.J.; Wolf, C.; Newsome, T.M.; Barnard, P.; Moomaw, W.R. World scientists' warning of a climate emergency. *Bioscience* **2019**, *70*, 8–12. [CrossRef]
37. Whyte, K. Our Ancestors' Dystopia Now: Indigenous Conservation and the Anthropocene. In *The Routledge Companion to the Environmental Humanities*; Heise, U., Christensen, J., Niemann, M., Eds.; Routledge: New York, NY, USA, 2017.
38. Harman, G. *Prince of Networks: Bruno Latour and Metaphysics*; re. Press: Prahran, Australia, 2010.
39. Harman, G. *Tool-Being: Heidegger and the Metaphysics of Objects*; Open Court: Peru, IL, USA, 2011.
40. Ulvila, M.; Wilén, K. Engaging with Plutocene: Moving Towards Degrowth and Post-Capitalistic Futures. In *Sustainability and Peaceful Coexistence for the Anthropocene*; Routledge: Abingdon, UK, 2017.
41. Heikkurinen, P. On the emergence of peaceful coexistence. In *Sustainability and Peaceful Coexistence for the Anthropocene*; Heikkurinen, P., Ed.; Routledge: Abingdon, UK, 2017; pp. 7–15.
42. NATIFS. The Souix Chef. Foundations of an Indigenous Food Systems Model. Available online: https://www.natifs.org/ (accessed on 15 November 2019).
43. Hunter, D.; Özkan, I.; Moura de Oliveira Beltrame, D.; Samarasinghe, W.L.G.; Wasike, V.W.; Charrondière, U.R.; Borelli, T.; Sokolow, J. Enabled or disabled: Is the environment right for using biodiversity to improve nutrition? *Front. Nutr.* **2016**, *3*, 14. [CrossRef]
44. Kuhnlein, H.V.; Receveur, O. Dietary change and traditional food systems of indigenous peoples. *Annu. Rev. Nutr.* **1996**, *16*, 417–442. [CrossRef]
45. *The Future of Food and Agriculture*; Food and Agriculture Organization of the United Nations: Rome, Italy, 2017.
46. McGregor, D. Honouring our Relations: An Anishnaabe Perspective. *Speak. Environ. Justice Can.* **2009**, *27*, 27–41.
47. McLeod, N. *Indigenous Poetics in Canada*; Wilfrid Laurier University Press: Waterloo, ON, Canada, 2014.
48. Clarkson, L.; Morrissette, V.; Régallet, G. *Our Responsibility to the Seventh Generation: Indigenous Peoples and Sustainable Development*; International Institute for Sustainable Development Winnipeg: Winnipeg, MB, Canada, 1992.
49. Robinson, T.D. Chronos and Kairos: Multiple futures and damaged consumption meaning. *Consum. Cult. Theory* **2015**, *17*, 129–154.

50. NASA/GISS (National Aeronautics and Space Administration/Goddard Institute for Space Studies). *Climate Data*; NASA/GISS: New York, NY, USA, 2018.
51. Barnosky, A.D.; Hadly, E.A.; Bascompte, J.; Berlow, E.L.; Brown, J.H.; Fortelius, M.; Getz, W.M.; Harte, J.; Hastings, A.; Marquet, P.A. Approaching a state shift in Earth's Biosphere. *Nature* **2012**, *486*, 52–58. [CrossRef]
52. Purser, R.E.; Montuori, A. Ecocentrism is in the eye of the beholder. *Acad. Manag. Rev.* **1996**, *21*, 611–613.
53. Purser, R.E.; Park, C.; Montuori, A. Limits to anthropocentrism: Toward an ecocentric organization paradigm? *Acad. Manag. Rev.* **1995**, *20*, 1053–1089. [CrossRef]
54. Hopwood, B.; Mellor, M.; O'Brien, G. Sustainable development: Mapping different approaches. *Sustain. Dev.* **2005**, *13*, 38–52. [CrossRef]
55. Rockström, J.; Steffen, W.; Noone, K.; Persson, Å; Chapin, F.S., III; Lambin, E.F.; Lenton, T.M.; Scheffer, M.; Folke, C.; Schellnhuber, H.J. A safe operating space for humanity. *Nature* **2009**, *461*, 472–475.
56. Popkin, B.M. Global nutrition dynamics: The world is shifting rapidly toward a diet linked with noncommunicable diseases. *Am. J. Clin. Nutr.* **2006**, *84*, 289–298. [CrossRef] [PubMed]
57. Popkin, B.M.; Adair, L.S.; Ng, S.W. Global nutrition transition and the pandemic of obesity in developing countries. *Nutr. Rev.* **2012**, *70*, 3–21. [CrossRef] [PubMed]
58. Mbow, C.; Rosenzweig, C.; Barioni, L.; Benton, T.; Herrero, M.; Krishnapillai, M.; Liwenga, E.; Pradhan, P.; Rivera-Ferre, M.; Sapkota, T.; et al. Food Security. In *Climate Change and Land*; IPCC: Geneva, Switzerland; World Meteorological Organization: Geneva, Switzerland, 2019; Chapter 5.
59. Tuomisto, H.L. The complexity of sustainable diets. *Nat. Ecol. Evol.* **2019**, *3*, 720–721. [CrossRef] [PubMed]
60. Parodi, A.; Leip, A.; De Boer, I.; Slegers, P.M.; Ziegler, F.; Temme, E.H.; Herrero, M.; Tuomisto, H.; Valin, H.; Van Middelaar, C.E. The potential of future foods for sustainable and healthy diets. *Nat. Sustain.* **2018**, *1*, 782–789. [CrossRef]
61. Ingram, J. A food systems approach to researching food security and its interactions with global environmental change. *Food Secur.* **2011**, *3*, 417–431. [CrossRef]
62. Schmidhuber, J.; Tubiello, F.N. Global food security under climate change. *Proc. Natl. Acad. Sci. USA* **2007**, *104*, 19703–19708. [CrossRef]
63. Naess, A. The shallow and the deep, long-range ecology movement. A summary. *Inquiry* **1973**, *16*, 95–100. [CrossRef]
64. Naess, A. *Ecology, Community and Lifestyle: Outline of an Ecosophy*; Cambridge University Press: Cambridge, UK, 1990.
65. Ingold, T.; Palsson, G. *Biosocial Becomings: Integrating Social and Biological Anthropology*; Cambridge University Press: Cambridge, UK, 2013.
66. Tuomisto, H.L.; Teixeira de Mattos, M.J. Environmental impacts of cultured meat production. *Environ. Sci. Technol.* **2011**, *45*, 6117–6123. [CrossRef] [PubMed]
67. Mattick, C.S.; Landis, A.E.; Allenby, B.R.; Genovese, N.J. Anticipatory life cycle analysis of in vitro biomass cultivation for cultured meat production in the united states. *Environ. Sci. Technol.* **2015**, *49*, 11941–11949. [CrossRef] [PubMed]
68. Lusk, J.L.; McCluskey, J. Understanding the impacts of food consumer choice and food policy outcomes. *Appl. Econ. Perspect. Policy* **2018**, *40*, 5–21. [CrossRef]

 *sustainability*

Article

From the Anthropocene to an 'Ecocene'—Eco-Phenomenological Perspectives on Embodied, Anthrodecentric Transformations towards Enlivening Practices of Organising Sustainably

Wendelin M. Küpers [1,2]

[1] ICN Business School ARTEM, 54000 Nancy, France; wendelin.kuepers@icn-artem.com
[2] Karlshochschule International University, Karslstrasse 36, 76133 Karlsruhe, Germany

Received: 27 January 2020; Accepted: 15 April 2020; Published: 1 May 2020

Abstract: The following paper discusses the contexts, conditions and implications of the so-called 'Anthropocene' (1). In particular, the following challenges the hyper-separation between nature and culture (2). Afterwards, possibilities for an anthro-decentric transformation are outlined (3). For this transformation-and following (eco) phenomenology-then the role of the body and embodiment, as well as a body-mediated turn towards an enlivening 'ecocene' is discussed (4). The article concludes with some implications and perspectives which are all related to a different kind of more sustainable organizing (5).

Keywords: Anthropocene; transformation; embodiment; organising; eco-phenomenology

---

1. Introduction

With its unprecedented changes in the earth's geo-and biosphere, the fundamental and irreversible human imprint and impact on natural systems and processes has turned humankind into a geological agent, which has led to term this epoch and state of affairs, the 'Anthropocene'. Under the techno-human condition [1], anthropogenic-induced environmental change and the domination of the Earth's ecosystems have reached a global scope and a permanent geological time-scale [2,3]. The detrimental impacts and degradation of human activities have accelerated and intensified in various areas. This includes the overuse of resources, and overloading that is over the limits, and by all this endangering the integrity of the fragile ecologies of planet Earth and its climate. [4].

Each of these and other human-bound phenomena are aspects of a tremendous upsurge in economic and other growth-related processes by of global productions and consumptions while being distributed via worldwide markets and globalization. These developments and their impacts are increasingly all-pervading in modes of hyper-modernist, neoliberal worldviews and its economic and socio-cultural practices. All of these activities and its impact are reaching and surpassing the boundaries of what the Earth, can cope or 'bear'. Exploiting the planet, as a resource-pool in unsustainable ways will lead to unforeseeable repercussions, side-effects and far-reaching consequences. In addition to passing globally aggregated earthly thresholds, infringing various planetary boundaries manifest in life-threatening impacts, such as climate change, which is more a heating and regional-level boundaries. These boundaries can be specified with regard to atmosphere and biosphere-integrity, geo-bio-chemical flows, land-system change, freshwater use, etc., while they have been crossed as a result of human activity.

Concretely, too much greenhouse gas going into the atmosphere, too much fertiliser running off into waterways and increased soil and ocean acidification, as well as overharvesting and overfishing are changing the organic makeup of planetary life. Furthermore, the implications of increased urbanization, contamination, losses of forests and wild-life; thus, bio-diversity, or species extinction, and further forms of human-caused destruction, have placed Earth's ecological environment and systems into a state that can no longer provide healthy and sustainable habitats. Even more, the complexities of today's interwoven, multifaceted problems that have arisen from the human domination of the planet call for a radical transformation.

Metaphorically speaking, the Earth is a body that is inflicted with serious ills. The earth-body's bio-spherical and socio-ecological metabolism cannot 'digest' human interferences, interventions and outputs. This greedy species called humans and its voracious mechanisms of capitalism are depleting what has been reduced to resources faster than what can be renewed, while emitting too many waste-products that cannot be absorbed. In fact, this effluence and pollution underline the 'weight' of humanity on the Earth. And thus this stress the way in which humans have become a collective, global subject' or in the words of Michel Serres: the "only thing left floating will be the homogeneous excrement of the victorious Great Owner, Sapiens sapiens" [4]. According to Serres, humanity has become the most successful parasite of any invasive species, a hominiscence appropriating the entire planet—taking without giving and engaging in continual (terrorizingly) territorialisation [5]. For him, this condition calls for learning to form a new natural contract to be woven into our existing sense of the social contract [6]. This would allow the recognition of Biogea, that is the formidable residual reality that keeps us alive, transcends us and can eradicate us [7] (p. 48). Such an approach invites the cultivation of an ecological 'ethics of symbiotic reciprocity with implications for reinventing, understanding, organising and managing as well as leadership education' [8].

One of the main drivers of this excessive matter/energy throughput is the construct of an unlimited economic growth in a limited world that is nihilistically producing ecological and other forms of devastation and destruction. In particular, it is an inherent 'productivist' orientation that is inherently unsustainable as is producing both burn-outs and over-shoots [9].

If we continue to transgress natural and planetary boundaries, of inhabitant of Earth namely humankind will cause the earth to destabilise, and constrain even more attempts to diminish existing suffering. All of the increasingly visible suffering and problems in ecological spheres mirror psychological and social alienations and pathologies. These problematical developments may lead to a decline and worsening of states of affairs in many parts of the world [10], as well as to an overarching threat to the entire planet and all its life-forms.

As the latest data and information about climate and emissions and the impact of carbon-based production and life-styles confirm, what will take place are very unsettling and uncontainable changes of the planetary climate and thus life on Earth. As we know, these radical changes will then generate impacts with disastrous effects on livelihoods. Part of the consequences may even entail migration, disease, civil conflicts, violence and war among humans and so-called non-humans alike.

However, processing the supposed inevitable collapse, and approaching probable catastrophe or possible extinction may service deconstructing, reframing and rendering creative responses of adaptation [11]. Or such movement towards an abys may mediate a radical rethinking and a reimagining a commitment to a more sustainable sense-making as well as very practically doing things and living differently [12] (p. 14).

In a way, the human constructed but unsustainable realities are demonstrating an existential precariousness of our way of living and as much a loss of control, as the need for recognition of what is non-controllable. In addition, "maybe the life-energy sprung by environmental and climate tragedy is to be found precisely in the liberation from one pattern of self-understanding here, into a vitally different one—a new sense of the human self in evaluative action with which climate tragedy challenges us" [13] (p. 12).

In the acute context we can experience an increased questioning of the quest for sustainability by the drama of the Anthropocene [14], sceneries of encountering it [15], or depoliticising and necropolitical 'Anthropo-obScene' [16,17]. These obscene 'anthropocenic' dramas and sceneries are threatening the entire planet earth and its forms of life which are al facing far-reaching impacts and effects while slipping into dangerous impasses.

Conceptual disconnection and practical alienation from materialities and bodies, the appropriation of (capitalist) desires, and fantasies of constant growth and repeatable progress have gained increasingly ideological traction and power. Disembedded from their material origin(ation)s, fossil fuels, like coal, gas and oil as well as other so-called raw-materials, function as energy resources that have ignited a Promethean fire. This has fired up first civilization and later an industrialisation machine, enframed in the illusory image of endless linearized growth. Over time, such orientation has generated not only a vast emptying out, but also devastating by-products or supposed side-effects that become increasingly a central problem. Reconstructing the narrative economic history shows how the carbon-centric development is based on replacing flow power (photosynthesis) and animate power, embodied in living creatures, with stock energy, based on fossil, and on capitalist accumulation. These developments have caused and are leading to a craving, polluting and inequality-generating 'capitalocenic' formation. This human-made 'capitalcenician' regime with its nihilistically productive operations, scientific knowledge and technologies, is generating noxious consequences e.g., of adding large quantities of carbon dioxide to the atmosphere and ecological devastation. Following neoliberal regimes and doctrines, current dominant systems of production, reproduction, and consumption as well as organisation and managerialism make it impossible for the 'earth-body's' bio-spherical and the socio-ecological metabolism to digest and bear human interferences, interventions and outputs in this very Anthropocene.

But entering the debate on the Anthropocene implies constant conceptual exchange between different planet-centred and human-centred 'hi-stories' that are increasingly being told and unfolding. These differentiating histories are concerning those between earth-system history (humans as geo-biophysical force) and world history (humans exerting social-existential power). Furthermore, the different stories are situated between a scientific life, involving measurements and debates among qualified scientists and a more popular life as a moral-political issue; also involve questions of culpability and responsibility. Accordingly, "there are many Anthropocene out there, used for different purposes along different lines of logic in different disciplines" [18] (p. 124).

As the Anthropocene is an ambivalent and malleable concept, it accommodates several co-existing and, at times contested, narratives and ambivalent visions. On the one hand, there are those gloomy dystopian and (post-)apocalyptic imaginaries with its stories of disaster, decline, demise and extinction. Correspondingly eco-eschatological narrative frames [19] and 'dark ecology' [20], are announcing the impending or actual arrival of the end of the world [21] (p. 7), as a revelatory climate Armageddon, as an example of many apocalyptic imaginaries.

On the other hand, there emerge neo- and eco-modernist visions that speak again about myths of progress, salvation and solutionist masteries that are supposed to be possible. The later ones are images of a heroic neo-Prometheian Earth-mastery that assume using techno-scientific knowledge and neo-technocratic expertise while pursuing a hyper-accelerationist vision. Such vision embraces not only 'big money', but also 'big science' with 'big pictures' of geo-engineering and 'big data' that all are seen as necessary to rescue the Earth and all 'Earthlings' [22], while being part of problematic agenda and a politics of hope that is partly illusionary [23].

Both apocalypticist and salvationist approaches follow questionable narrative trajectories that are often deployed as discursive-representational tactics or rhetorical strategies. Such imaginaries and interpretations are often communicated by alarming Cassandrian advocates, who are stating that all is doomed, versus those Panglossian advocates according to whom all is supposed to be the 'best' in this best of all possible worlds. Both agendas, - those of 'dooms-dayers', who try to slow all down or resign while getting ready to the last days, as much as those 'boomsters', who call for all to 'speed it up' in an practicalism - both of these are problematic, if taken as one-sided or fundamentalist. In their, place- and timeless abstract positioning, these dystopian and utopian orientations are foreclosing political debate. Such, reductive imaginaries are making it harder to ask critical questions, like how to get out of reproductive, unsustainable socio-environmental trajectories of organising and managing.

We are facing an exhausted or fatigued quest for sustainability as an eco-political paradigm that is merely sustaining the unsustainable [24]. Accordingly, there are various discourses about an 'end of sustainability' [25] or 'post-sustainability' [26] while also calls for alternative forms of post-capitalism, post-growth and post-consumerism emerge. Co-responding to the given situation and the outlined discourses and calls, the following offers some perspectives towards mind- and artful ways of becoming sustainable with and as cultures of complexity [27,28].

Instead of seeing the Anthropocene as a single hegemonic, linear narrative that is supposedly apolitical, the subsequent elaborations recognise multiple, debatable or controversial and even polemical narratives [29]. Informed by deep insights from past histories, decentred, post-anthropocenic imaginaries may then help reframing existing relation to the present, while inviting and encouraging more creative responses and promising alternative sustainable futures to come.

While the meaning of the age of the anthropos and its various ways of being narrated and researched remain disputed and unsettled, there is a distinct and particularly relevant plotted storyline and quest emerging. This patterned emplotment concerns the relation between nature and human/culture. It seem that the juxta-positioned relationship between them will be a question of life and/or death-at the border of earth-history [30]—as its impacts generate unsustainable realities for both as a meshed nexus.

The overcoming of dichotomies as separating oppositions and leaving behind apocalyptical dystopia [31] or bleak optimism [32]—will be required in order to move towards anthropo-decentric transformational futures. These separating dichotomies between nature and man/culture may also be connected to opposition between human and non-human, structure and agency, micro and macro or surface and depth.

Misleading dichotomies are behind the logic and arguments for the Anthropocene. In particular, what is presupposed is that a claimed exceptional human reality is not a geo-logical reality and that humans are of more value than the ecologies in which these are embedded. The dividing understanding of humans as extra-natural and nature as extra-human, has triggered and reproduced the very stage for a systemic appropriation and exploitation.

Based on all the outlined critique the next sections first challenges the hyper-separation between nature and culture and then outlines possibilities for an anthro-decentric transformation. With regard to the latter, the author will then address the role of a body-mediated turn towards an enlivening 'ecocene' and concludes by offering some implications and possible perspectives.

2. Overcoming the Hyper-Separation between Nature and Man/Culture

Ontologically, epistemologically and practically, the Anthropocene challenges the traditional distinctions that are separating nature from culture that is from cultural structures, and the order of approaches and knowledge about the world and social practices.

As the increasing number of Anthropocene-related analyses is showing, humankind has monstrously scaled up in forces with a global expansion. These forces demonstrate the acute capability to radically reshaping the Earth. In this very era of the Anthropos, "natural forces and human forces are so intertwined that the fate of one determines the fate of the other" [33] (p. 2231). Such entwinement of human and non-human spheres and socio-natural entangled hybridization in a climate-changed world [34] has led to what has been called a post-natural ontology of the Anthropocene [35]. Such ontology is characterised by crossing the human–nature divide inherited, especially in the Western world from Platonian–Cartesian–Newtonian paradigms and the Enlightenment and Modern eras, as foundations of 'humanism'.

Separating the natural and the human has led to the functioning of an anthropological machine [36] by which animals are 'humanised' and humans are 'animalised'. As Agamben showed, the operations of anthro-machines gradually lead to immoral dehumanising acts and a mechanization of embodied life that implies the loss of meaning. For Agamben, these developments cause a valorisation not of 'bare life' but of efficiency, along with the exchangeability and disposability of all that can be assembled or disassembled.

Nature has been and is domesticated, technologised and capitalised in a way that it even cannot any longer be considered as what was used to be called 'natural'. Whereas nature is 'humanised' in the sense of anthropo- and socio-genic practices, these same practices are normalised or 'naturalised' and thus understood as part of 'natural' occurrences.

As epoch and as discourse, the Anthropocene appears as two entwined processes and conceptualisations. On the one hand, it is happening in real time; that is, it is chronologically characterised by the human bound impact on natural environments in the context of historical time. On the other hand, it represents given states of relationships between humanity and its societies, positioning nature as that which is at stake, whereas the supposed separation from nature supports our perception of entanglement.

Realising humanity's material dependence, embodiment, and the fragility of beings including human ones calls for rethinking and reimagining those traditional assumptions and myth about the autonomy-based self-contained and rational subject that commences and terminates with itself [37]. This state of affairs demands a decentring of the anthropocentric dispositions and positioning, and critically problematising tendencies towards the re-elevating of humans as reborn Prometheuses [23,38]. These neo-Promethean Earth-master are now "playing God with the climate" [39], while rivalling the great forces of nature of a defiant earth [40] (p. 41) by pursuing an updated eco-modernist agenda. Such agenda would continue and perpetuate the human-centric, technocratic managerial program that is characterised by hyper-accelerationist vision and actions to save both planet earth and its earthlings [41].

By contrast, what is needed is a critical understanding of anthrogenic and androgenic orientations that favour an eco-modernising and its implications, as well as eco-authoritative claims [42]. This need for such a critique is even more urgent, as it is emerging from different socio-political narratives and 'bio-power(ful)' settings that produce vulnerabilities and inequalities as well as unintended consequences.

On the one hand, and considering environmental problems and inequities, it is understandable for humans to debate their technocratic options: the speed of transition to renewable energy, issues of justice in relation to the climate, possibilities to sequester carbon, harvesting rainwater, securing food, developing policies concerning climate refugees, measures for adaptation and mitigation etc. On the other hand, these supposed solutions are part of the problem: since if continued in a Promethean manner, such 'solutions' will generate further problematic im- and complications as well as further possibly unsustainable developments.

One of the problems with de-politicising eco-modernist approaches is an immunological, bio-political fantasy that sustains the promise of adaptive terraforming, and resilience. Such imaginary represents a management of earth systems that supposed to allow perpetuating known life-forms for some, while becoming a deadening politics, or 'necro-politics' for others [13].

The challenge will be developing a different alternative approach. This would be one that processes differences sensitively, and simultaneously is more inclusive concerning the very status of the entangled material-based and the human-mediated spheres seen in a newly understood and enacted continuum of the natural and cultural life and its worlds. These life-worlds are those of embodied and socio-culturally 'constructed' institutions and organizations that are composed and immersed or consuming physical matter and so-called non-human dimensions. They constitute entanglements of so-called 'non-human' materiality and social 'culturality'. These worlds of life can be reinterpreted together as a kind of 'materio-culture' as well as 'socio-nature'. Such nexus of 'materio-cum-culture' and 'socio-natural' relations [43] refer to an entwining mesh. Interrelationally this mesh is forming a pluralistic and complex process that is unfolding as an uneven, contingent and emergent process.

Correspondingly, at present there exist various strands of neo-materialism and critical post-humanism that reflect this meshed nexus of culture and nature as confluent 'natureculture' [44] or entwined 'nature-culture' [45]. These diverse orientations aim at a non- or post-anthropocentric approaches and mappings. All of them search for an understanding of matter and what matters, socio-culturally as well as reflect 'matters of concern' related to generative and democratic powers. Such interpretations allow conceptualizations of wayfaring movements of confluent fluxes *in-between* matter and minds, bodies and souls, natures and cultures etc. Entering these 'in-betweens' may open up imaginations and envisioning, while entailing far-reaching 'real-world' immersion and engagements that relate to disclosing phenomena and opening practices that reverse established divisions between what was used to be called 'nature' and 'culture' as separate constructs. Both are reversed and repositioned into a what can be expressed in a new synthesizing word as a 'nat-cult-ural' integration, for example, related to organization and leadership [45].

3. Anthro-Decentric Transformations through Embodiment and Bodied Practice

The interweaving of distributed and 'performing' materialities and 'other-than-human' as well as socio-cultural realities infer and refer to a specific immanence. It is an immanent sphere that can be seen as a form of 'living mattering' of entangled agencies and expressing symbolic and political meanings. Importantly, this immanent materiality and mattering is always already constituted and mediated by bodies and forms of embodiment. The sensual and social dimensions of the material and 'post-human', play a constitutive role in the structuring and organising of situations and configurations as enacted within various life-worlds. These worlds of life— also within organizations and leadership [46]—are always already part of an embodied performing processes in and through organising [47].

What is called for are ways to facilitate the processing of multiple futures of Anthropocenes, beyond an age of humans [48]. This idea is even more 'applicable', as the 'anthropos' has never been only human' [49]. An abstract human or humanity as homogenous unit or 'collective' actor that is completely dependent upon nature never existed, and the supposed human exceptionalism is a constructed myth and misleading belief. The danger of such orientations is to frame and reduce human-based activities in the web of life to an abstract humanity as homogenous unit or as a 'collective' actor [50]. Critically we can ask: was it that all human-kind caused the Anthropocene or, rather more specific ways of living and organising, in particular ways of managing and politics, as well as definite forms of technologies and its usages that caused profound alterations of life and ecology?

Instead of glossing over with universalising tendencies, differentiate among the multiple and unequal social values, relations, and practices of power that accompany 'essentialised' humans and their individual and collective behaviours. This is true, especially when materio-natural phenomena and environmental changes are related to social-cultural categories and realities, such as classes, races, genders, powers, or capitals etc.

In particular, the aforementioned eco-modernist orientation can be a misleading 'story' or ideology, according to which 'human enterprise is contrasted with the great forces of nature [3,51]. This contrasting logic follows the idea that 'Humans' are part of nature, whereas 'Humanity' is separated from 'Nature'. Considering these oppositional dis-positions and that one key critique of Anthropocene-talks is the very pitting of Man against Nature 'out there', critical questions emerge:

- How can we extricate ourselves from this schizoid interpretation of defining humanity as a species within an ecology of the living, with regard to methodologies, strategies of analysis, and structures and ways of communicating, human activities are seen and treated as separate and independent?
- How do we relinquish Cartesian dualism and leave behind or process dichotomies and separations between the ecological insight of a 'humanity-in-nature' versus the ongoing dissection between humanity and nature? How can we bridge the 'Great Divide' [52] between the two? Or, do we need an 'Ecology of Separation'? [41]. Acknowledging the wild, subtractive capacity of nature, and against eco-modernist technocratic delusions or conservative 'organicism', can there exists an unconstructable Earth that emerges as unsubstitutable becoming that always escapes the hubris of those, who would remake and master [22]?

- How can we 'save' nature in a 'post-wild world' [53]? How can we sense and make sense of another means of materio-socio-culturally being and becoming in the world, where the social and other categories are not only human [54]?
- Can the Anthropocene epoch be one of brave experiments and a deep regeneration of modes of how to organise and manage, including giving presence and expression as well as agencies or legal forms of prudence and protections to human and other-then human beings [55] (p. 13)?

With regard to anthropocentrism, humanity seems to be both "Peril and Promise" [56], while questions about the status of intrinsic value of non-human or more-than-human beings are vital. The challenge will be to overcome one-sided subjectivist and objectivist definitions of value and valuing. With the subjectivist, human-centred approach, all attributions of intrinsic value are anthropogenic, that is, originating in and dependent upon human acts of evaluation [57]. By contrast, for an objectivist environmental holism, pursuing an ethical non-anthropocentrism, an ecosystem possesses a systemic value e.g., 'producing' life. Such system-based value is seen as both fundamental and prior to the intrinsic value of living things, including human beings and exists independently of either human evaluative or perceptive acts [58,59].

Value is neither just a human social construct, nor a free creation of human desire, labour or capital. Nor do other-than-humans only have a system or ecological value, but both are media and agents of value creation. How creative and flexible can then our human value perspective be regarding the other-than-human world without breaching an objectionable anthropomorphism?

Enlightened anthropocentrism and various forms of non-anthropocentrism, sentio-centrism, biocentrism, and ecocentrism are still centred on human versus 'non-human' speciesism. However, as outlined what is called for is a decentred and decentring, inter-relational approach that is neither 'subjective', nor 'objective'. Rather such a radical inter-relational understanding and practice requires new terms —like *'inter-jective'* and *'inter-jectivity'*—that allow a sustainable flourishing in what will be called in the following the era of an 'ecocene '. The subsequent section will discuss how an eco-phenomenological perspective on embodiment can be helpful for qualifying that is perceiving, comprehending and enacting this 'ecocene'.

4. Merleau-Pontian Phenomenology for and in the 'Ecocene'

The phenomenology and late ontology of Merleau-Ponty provides a decentring post-Cartesiam-centered approach for developing not only a timely environmental ethics [60], but also 'proto-wise' practices for an co-scene.What makes Merleau-Ponty's phenomenological approach towards the body and of embodiment so valuable for a post-anthropocentric understanding is his radical criticism of one-sided forms of materialism and idealism. Following a certain logical exclusion, both reduce living phenomena and ways of sensing and perceiving them either to the realm of matter or to that of ideas. With this both of them are each failing to have access and understand expressive senses of qualities and practices emerging into what Merleau-Ponty calls embodied 'inter-becoming' [45,61].

Beyond reductive materialistic or idealistic positioning, he develops an understanding of practising that is mediated by bodies and being facets of a post-dichotomous nexus of "self-other-things" [62] (p. 57). In turn, this is part of a perspectival and transformative "integral being" [63] (p. 84). Living bodies serve here as media for moving into post-anthropocentric 'ecocene's. In these eco-scenic spheres the nexus of mind-and-matter, culture-and-nature, self = and-world, are gathering and co-evolving together in non-fusing but holonic ways of part and wholes, thereby disclose an open process of 'betweening'. The embodied 'subject' or 'inter-(sub)ject' and its likewise embodied 'inter(sub)jective' encounters with all their meanings and 'inter-(ob)jective' lifeworld and its forces are an extensive continuum. In the same and different ways, all of them are rooted and co-create together while taking part in a mode of 'inter-being' [64] (p. 208).

As uncontrollable, erratic, and unmanageable be(com)ings, the body and embodiment are genuinely decentring. Neither is centred or mastering, but rather disrupting, dejecting, and absconding purposive and boundary-defining impositions or orders. Bodies and embodied forces cause dynamic and incomplete ways to perceive, feel, think, intend or respond as well as to act om material, biological, social and cultural spheres. All of these dimensions are twistingly entwined and jointly engaged within an ever-present relational sphere. This special sphere is what Merleau-Ponty named the 'flesh' of the world [63]. In particular this flesh can be characterised as an elemental carnality and formative medium for an in-between that is relevant for all processual ways of organising [47], reinterpreted as 'organic-isations' [65].

The philosophy of nature by Merleau-Ponty [66] contributes to the development of a responsible ecology [67,68] and an affective eco-phenomenology [60,69]. These orientations are both elements of a reawakening of corporeal planetary senses and an invigorated dance of entangled 'Earth-bodies' [70]. Such planetary bodies move in co-emerging, spiraling ways, while embodying, and unfolding the 'human-and-other-human' world, and being attentive to how silence sings the world [71].

A Merleau-Pontian phenomenologists approach does take seriously the values and contributions of various beings on their own terms and in relation to the 'more-than-human' beings in their otherness [72] (p. 48). Through our embodiment, we can learn to acknowledge relations of momentous ontological continuity with other beings, including animals, plants and, (a)biota.

The charge of an Anthropocentrism against Merleau-Ponty fails to recognise that as a body-subject, one is not solely, nor even fully 'human', as one's body implicates one *in* the world also comprise others within [72] (p. 54).

Bodies are living, and dynamic affective openings that take part in a dis- and continuous process of relating and meaning-making or releasement of something meaningful. These are bodies with rhythms and pulsations that affectively engage with, withdraw from and reconnect again with the world and its phenomenal fields. These bodies exist together as 'human-other-human' conviviality that is as being with a natural ('con-naturality') and a material-discursive conviviality as being with the social ('con-sociability'). Expressed in terms of material ecocriticism in both of these forms together, there is 'storied matter'. This narrative materiality speaks to or about the ways that material forms—bodies, things, elements of organic as well as inorganic matter, landscapes, and biological entities—"intra-act with each other and with the human dimension, producing configurations of meanings and discourses that we can interpret as stories" [73] (p. 7).

This intra-actional process is possible due to an evolutionary continuity that is an ontological crossover via lateral kinship with other bodies [63] (pp. 207–208). Moreover, based on this and through a genuine conviviality, a coevolving symbiogensic 'Ecozoic' era [74] may emerge. As the 'con-naturalities' and 'con-sociabilities'—present in conjunction with convivialities—are not directed towards a teleological convergence, both and moves towards that ecozoic world keep a momentum of living differences that is and remains open for other un- and re-foldments or foldings. Cultivating and establishing new ecozoic or ecocenic' relationships between human- and 'more-than-human' beings will require disruption of assumptions of both hierarchies of importance and centrality that are positioning humans in the centre and move towards a mutual thriving.

The challenge will be to develop more comprehensive understandings of entwined 'wilderness' and 'culturedness'. Such understanding is one that honours the presence and voice of 'more-or-other-than-(human)-beings' and even more expressions of multispecies, in multi-de-centric and convivial ways. With an ontological relationism, we can cultivate relearning to sense, listen and otherwise perceive earthly phenomena and develop responsive attending and caring relationships. Such attending and caring is considering to what is going on within and around in a continual emergence in relation to worldly conviviality across vast webs of difference beyond narrow personal and human-centred concerns, and organizing a transformation towards a living 'ecocene'. Thus, overall Merleau-Ponty's eco-phenomenology and ontology can help to develop transformative commitments and forms of organising for more sustainable futures [75]. What is needed to enact such futures is a revived re-thinking, re-imagining and re-doing as well as cultivating proto-wise, pro-sustainable practices that all may serve as media for developing and organising anthro-decentric orientations.

Questioning the anthropos of the Anthropocene, such orientation allows not only to decentre and re-situate the anthropos in relation to the earth. Rather, such undertaking is envisioning a new understanding and enacting of 'Gaia' as a geo-bio-historical multitude and assemblage of coevolving life-forms. From such decentring and interrelational Gaia-ian perspective Gaia the anthropos appears then as what could be called 'humanimality' or 'hum-animism' [76] as part of the above-described entwined 'nature-culture' These 'humanimals' are natural-cultural beings of Gaia understood as a multiplicity or distributed agency as Latour has discussed in his enquiry into natural religion and new geo- and ecostories [77]. Not having fixated intentionalities, eco-stories have their starting point in the inherent force of agency. Thus, they can be defined as "a form of narration inside which all the former props and passive agents have become active without, for that, being part of a giant plot written by some overseeing entity" [77] (p. 74). Such decentring nexus can be interpreted as 'materio-culture' [78] as well as 'socio-nature' [79]. Both of these terms refer to a nexus of entwined material, social, and cultural ecologies as embodied, relational practices of human and other beings in relation to responsive and responsible organizations [80].

A critical usage of 'anthropocenian' concepts provides a chance for developing different, more embodied, understandings, imaginaries [81] and practices of 'ecocening'. These concepts concern not only multi-speci(es)fied arrangements and ecological embodiments, but also embodied, prudent sense-makings and practical 'proto-wise' imaginaries. Such 'alter-Anthropocene imaginaries' [82] express stories, policies, actions or 'non-actions', and corresponding forms of governance, institutions organizations and their leadership. Furthermore, an 'ecocening' can help to challenge imperatives of progress and growth as well as further belief-systems that underpin economic, organisational, and managerial thinking, constructing, and acting.

Taking an embodied perspective, facing the impasses of Anthropocene 'ecocening', understood also as putting living ecologies 'on the scene', has the potential to usher into a radical transformation. In such transformation an appreciation of the nature and culture, both 'materio~culturally' and socio~naturally with its 'impActing' and extent of human impacts on global environmental processes can be a leveraging medium or catalyst for an imaginary and practical shifts. Such transformative shift includes a radical change in affective and cognitive dispositions, and attitudes, thus ways of perceiving and conceiving as well as cultivating different habits and habitus in relation to the earthly habitat. In other words, such transformational realisation facilitates the emergence of sensibilities and sense-making practices in and through organizations that are of the magnitude required to learn living and working differently, individually, collectively and systemically. Moreover, an extended organising would involve also playful 'kin-making with other-than-human beings we share this planet with.

Thereby, the Anthropocene and its impasse may serve to inter- and disrupt as well as go beyond the limits of conventional, inherited worldviews, forms of knowing and its enactments. Questioning the same can then contribute to re-visioning and re-enacting individual and social customs of sensing, thinking, knowing and doing.

A caring imagination, especially an ecological imagination [82] or imagining anew may allow rendering the narrative ethical and aesthetic dimensions sense-able thus morally motivating [83] (p. 150). This rendering mediates perceiving actual anthroposcenic conditions as visible in light of what could be eco-scenically possible.

A corresponding social dreaming serves as a creative source and a corresponding eco-scenic ethical practice [84]. Similar like a creative moral imagination in an age of globalization [85–87], an eco-ecenic imagination helps to re-orient and reframe organizational and managerial practices, collective actions, more moral 'outcomes' [88] and ethical decision making. These 'proto-ecocenic' imaginations can be seen as situated and creative activities in relation to anthropocenic problems and dilemmas. Forces of imagination draw together different possibilities, freeing them by suspending between determinate and indeterminate vectors of sense. This is allowing for deliberate - de-liberarare' means weighting out - risks and opportunities, and creating possibilities for emerging processing and actions to 'take place'.

Following Bachelard's phenomenology of material and elemental imagination and ontology of imaginative creation in waking-working reveries [89], these could represent the groping experimentation of 'imagining forward'. Furthermore, these are forces of becoming that can help reconcile man and nature as an 'inter-world' of trans-subjective reveries and poetic images for the 21st century [90]. Such attempt at re-imagining and 're-story-ation' seeks to rediscover an elemental connection with earth, fire, air and water, that allows a felt as well as imagined poetics of vital agency of matter and 'proto-environmentalist' practice As part of rediscovering our planetary senses, for Mazis [70,71], imagination can serve as a medium through which the symbolic accumulations of different past moments are connected and revealed. And this can happen in a way that makes them resonate with one another in the light of the present and multiple possible futures. In particular, a poetic turn—resonating with a sensitivity to silence and gestures—can help in drawing upon the imagination inherent in the perception of the physiognomy of the world. Moreover, such poeticising can serve as an entry into the depths of perception and the nexus of genuine INTER-relations among beings.

5. Turning towards an Enlivening Eco- or Zoë-cene and Ecological Practical Wisdom

Is a reversal possible, a seemingly unthinkable „Great Derangement"[91] of practices and life possible and what would such move mean and imply?

A post-anthroposcenic turn [34] might help re-discovering old forgotten ways or mediate new ways of being and becoming in the world meaningfully. A revived turning may then contribute to a transformative transition away from an overly human-centred, deadening Anthropocene towards a more inclusive, enlivening [92,93] 'ecocene' or 'zoë-cene'. Concerning the latter term, the Greek word *zoë* refers to meanings of life in its felt sense, including the whole animated and animating Earth [93]. Accordingly, the living zoë can be conceived as a generative and vitalist force that is common to all species in transversal domains [94]. An orientation towards such zoë-via a 'Gaian Praxecology' [95] integrates and learns from the proto-wisdom of plants and animals in the sense of an integral eco-sophy and human wisdom.

Eco-phrónêsis manifests an ecological practical wisdom for and from ecological practice, including planning, design, construction and management [96]. For such comprehensive wisdom perspectives as embodied practice, bodies and embodiments are entangled with worlds that are both natural and cultural. Thus, there is a "relationship of 'inter-Corporeity' with the biosphere and all animality" [64] (pp. 334–335).

Ecologically wise approaches sometimes favour an anthropoharmonism and anthropo-harmonic compassionate ecology [97–99]. The do so to work both actively and cooperatively with the wider biotic community to preserve, regenerate, and adapt healthy, functional, and resilient ecosystems, e.g., anthropoharmonic phrónêsis, such as the practical wisdom of permaculture [99].

Moving towards a re-embodied scene of zoë-living represents a tremendous opportunity to engage more wisely and transformatively with the quests and questions of ecology and its inhabitants as well as in and through organizations.

Facing increasingly alienating times of rapid and escalating change [100] and technical and social acceleration [101], such quests and questions involve the reconsideration and reintegration of meanings, values and responsibilities. All of this is as part of a complete reorientation of the human-earth relationship as well as its resonances [102].

The eco- and zoë-ianly qualified (post-)Anthroposcen-ian move marks the transforming of Earth—twisted by practices of the anthropos—that is radically altering organizations and society on a meso-level as well as their acting members on a micro-level towards wiser ways of operating and living.

Situating wisdom and its learning call for a reconfiguration of relationships among all beings with regard to local and planetary perspectives and by organising differently. Cultivating and enacting practical wisdom for this 'eco- and zoë-scene' is part of living well, i.e., individually, societally, ecologically and sustainably flourishing [103,104]. Realising this well-be(com)ing has never been more important than it is today and for a liveable world to come. Its urgency also affects the actual and possible, not-yet-incarnated lives of stakeholders that are at stake or, even more, whose stakes are already burning. Facing these auto-destructive dynamics, and he limiting character of the present situation of demand a metamorphosis beyond 'over-coming' and different ways of organising.

6. Organising (in/post) Anthropocene

Anthropocene has been and increasingly is discussed in the academic organization literature e.g., [55,105,106]. Very early, the anthropocentric orientation in organizations as 'a primary instrument by which humans impact their natural environment' [107] (p. 705) had been problematised [108]. The complex relationship between organization and the environment (al) crisis has invited eco-social reflection and inquiry [109]. Accordingly, calls have been made for greening organizational studies (e.g., [107]) and for eco-centric [108] and/or climate resilient organizations [110].

But then what does organising mean in the Anthropocene [106,111]? How can the Anthropocene help not only in understanding on-going anthropogenic problems and organising its solutions, but also creating alternative forms of organising based on other relationship between humans and more-then-human earthlings? And how can the emerging-geo-story of the Anthropocene contribute to and be translated to renewed local understandings and practices?

The conceptual and methodological challenge in responding to such questions lies in reflecting on how to connect geo-biophysical and geo-political concerns on the macro- and mundo-levels to the meso-levels of organising, as well as the micro levels of practitioners and individual actions. Furthermore, how to interrelate the temporal scales ranging between intergenerational and bio-geo-logical time to short- and mid-term orientations of organizations?

The overriding discourses and practices in organization with regard to environmental sustainability seems to offer no satisfactory or acceptable responses challenges of the Anthropocene that require epistemic shift, as long as no new 'matters of concern' [105] (p. 10), ecologies of concern [112] or responsible ecological concerns are taken seriously. Moreover, from feminist ecological perspectives, Seray et al. call for a move from a matter-of-fact orientation, towards an "aesthetic of matters of concern as necessary for reclaiming sustainability in organization theory for life in the Anthropocene" [105] (p. 10).

7. Practical, Political and Theoretical Implications

7.1. Practical Implications

The outlined dimensions and transformational moves from a deadening Anthroposcene towards an enlivening 'ecocening' have several practical and political, as well as theoretical and methodological, implications. Instead of being directly designed, embodied and enfleshed practice can only be designed for, that is, allowed and encouraged, to unfold [113]. Part of this challenging transformation is to provide supportive conditions and constructive relationships that engender targeted facilitations or catalytic circumstances on a situation-specific basis. By thes interventions embodied sustainable eco-scenic practices can then flourish in every-day work-life. Possibilities that enable and empower need to be organised in adapted and practically tailored ways. Specifically, this organising and offering has to be done according to the needs, demands and requirements of given environments, states of affairs or objectives aspired by employees, management, organizations and external stakeholders.

Enfleshed eco-scenic practices of sustainable development in embodied organisations [114] can bring to the fore, for example, concrete forms of energy-consumption, recycling, transport and food-practices concerning natural and social ecologies. This need to be related in particular to environmental workplace behaviours [115] or life-affirming, inclusive approaches towards biomimicry [116] and the rise of the biophilic organization [117]. A corresponding sustainable capability approach acknowledges the intrinsic value of other-than humans, including plants, animals and various ecological life-forms [118,119]. All of them need to be recognised as primordial stakeholders of Gaia [120] in an open-ended participation processes. What is practically required are forms of nurturing eco-cultures that mediate sustainable ways of living [16] and integral wisdom practices [121]. The challenge will be to change routined path-dependent practices, which means identifying possibilities for disrupting, reorienting, and/or redirecting practices in more sustainable forms of realisations [122,123], Importantly, this implies also understanding and overcoming psychological-cognitive and organizational obstacles to action that perpetuate unsustainable practices [124] and addressing political dimensions.

7.2. Political Implications

Organising differently in the Anthropocene and towards a more integral ecocene necessitates an ethico-political restructuring and transformation of contemporary organizations. This concerns, especially ways to support employees, groups and stakeholder to be able to engender proper ways to negotiate and respond as well as cultivate actual and proactive ways of acting that are more sustainable. Critically, this also involves analysing and questioning anthropocentric and interest-centric political practices and how they are used to accomplish and uphold power or control.

How are specific 'proto-eco-scenic' experiences, meanings and practices discriminated, marginalised, degraded and ignored, or how are attempts to enact them dominated, subordinated or disciplined? Correspondingly, a critical approach can be used for studying the ordering and normalising of disciplinary techniques (governmentalities in sensu Foucault) and encumbering processes of forced or imposed anthroposcenic affirmative practices. Furthermore, such an approach could explore how the dynamics of power and distress increase insensitivity to the pain of others [125].

In the same vein, another focus might be to study what and how practices are silenced or invisibilised in organizational regimes. This implies exploring what is not sensed, expressed or practised, including un-noted feelings, thoughts, actions or actors and omissions in the hierarchy of activities underpinning social organizational or managerial and economic orders. Additionally, such orienation suggests considering what seem to be strategically unthinkable, supposedly un-doable or tabooed, for example in making decisions and those ideas that are excluded as supposedly impossible practices.

A promising approach in this context is to consider the Anthropocene through the lens of Rancière [55,126–129], with his focus on emancipatory politics and social change—the order of the police and politics as well as the redistribution of the sensible. Following his post-foundational and postliberal democratic understanding of a disruptive politics of dissensus, this includes a re-arrangement of political order and different regimes of perceptual part-taking. These regimes regulate what can count as perception, experience or sense, not only individually, but also collectively.

The reconfiguration of these regimes modifies a sensory framework that distinguishes the visible from the invisible, the sayable from the unsayable, the audible from the inaudible, and the possible from the impossible with regard to ethical issues. In particular, an everyday aesthetics that affects the perception of and enables the redistribution of the sensible can be part of the reconfiguring of the common field of aesthetics and politics [130].

Future research on aesthetic ways of organising may comprise exploring the 'undoings' of a given distribution of the sensible, or re-orderings of the very manner in which (public) space lends itself to and is produced by processes of organising extended towards the general forces of organising that affect living beings and the conditions of lives [131] (p. 38).

This also involves instituting in the sense of establishing and asserting new political 'subjects' or better to say 'inter-jects' that are otherwise not acknowledged by the order of the policing powers. The new performing political subjects or 'inter-jects' can be linked to a theatrical and literally spatial dimensions of 'staging' that are implying a re-construction of the stage, the production of scenes, provoking performative places and times as well as moving bodies and voices or expressions that have been theretofore unheard of and unseen [132].

The inter-face of the previously outlined nexus of 'nature-culture' can be interpreted as a milieu where experiences, practices, policies, ideas and knowledge meet, are negotiated, discussed and resolved [133]. However, such an approach must consider also the ambivalence, tensions and problems involved in such undertaking. In particular such approach needs to avoid falling into a retro-romantic orientation, that is nostalgically pursuing or reverting towards a holistic harmony-seeking and sentimentally re-enchanting eco-philosophy.

As much as a certain longing for returning to a pre-reflective unity for disembodied, alienated humans in late modernity, or to a fragmented as relativistic postmodernism consciousness appears as understandable, there is no nostalgic way back. There is no way anymore towards retro-regressive coincidence with nature or supposed pre-existing given identities or 'Truth'. Because the reversibilities of being are always imminent and never realised completely, 'the coincidence eclipses at the moment of realization' [63] (p. 147). This implies that all relations to nature and to bodies are always already culturally mediated, to the extent that culture is 'natural' and embodied.

7.3. Theoretical and Research Implications

Regarding future research, theory building and empirical investigation, research can focus on the conditions and qualities of the transformation towards more eco-scenic organising and its embodied practices. For exploring this methodologically, a processual approach is appropriate that enacts the literal meaning of method as 'following along a way', i.e., 'meta ton hodon'. By disclosing descriptions and interpretations of actual experiences and phenomena as they appear, researchers can develop a much-called for a-causal, non-reductionistic and non-reifying approach towards ecocenic organising. In particular, longitudinal studies and multiple case studies as well as sensually oriented and art-based methods are suitable for what might be called 'pheno-pragmatic' research practices [47] (p. 253).

Future research on integral organization and leadership would be supplemented by exploring bodily mediated spatio-temporal and cultural realities and tacit experiences and knowing by using art-based research. This includes also sensually oriented methodologies and, for example, aesthetic ethnographies and interpretations [134]. Research, especially on embodied organising and learning, is "fully alive and creative when wide-eyed and involved, when it sees, touches, hears, tastes, and feels" [135] (p. 19), thus when it is using and refining embodied, sensory faculties. Therefore, considering the own body and other bodies while researching qualitatively [136], and various manifestations of embodiment in research [137], as well as deploying embodied research methods [138], is vital.

8. Conclusions

Critically, this contribution reflected upon questions of organising in the Anthropocene by discussing possibilities for overcoming the hyper-separation between nature and man/culture as well as anthro-decentric transformations. Outlining the need for turning wisely towards an enlivening eco- or zoë-cene, were complemented by offering practical, political and theoretical implications.

For exploring modes of togetherness and modi co-vivendi (conviviality), Merleau-Ponty's phenomenology of a non-dual, and non-anthropocentric flesh—as elemental, reversible chiasmic 'one-in-an-other' or 'Ineinander' [64] (p. 306)—calls for a more inclusive understanding of nature-and-culture. Such comprehension accommodates both human and 'other-than-human' components as part of a primordial, enfleshed in-division and to describe how various others like animals, plants and, albeit to a lesser extent, (a)biota, are and become alive by being related to each other.

Transforming from the Anthropocene to an ecocene and with this moving towards a "cosmoplocene" [138]–invites non-anthropocentric stories of the uni- or pluriverse. For some, this is part of an ex-centric anthropocosmic turn and narrative via the hovering forces of elemental imagination [139] (p. 172) to be eco-scenically processed in a critical way.

This transformation towards an ecocene may include an eco-sensitive, cosmopolitan, environmental narrative and wisdom-oriented ethics to cope with the globalised nature-culture of risk in the Anthropocene. As such, this evolution may facilitate a wiser move towards the thriving of other-and human beings. This would be a type of ecological eudaimonism [140] that comprises living well, i.e., individual, societal, ecological, and cosmic flourishing.

Ecocentric theorising and practice means leaving behind an unsustainable productivism [9], while cultivating an non-anthropocentric ethos of releasement. Such releasement-orientation allows human and non-or other-than-human worlds to peacefully coexist, co-evolve and co-creatively unfold on each on their own terms and in a new understood common sense related to releasement [141]. Such a mindful ethos of releasement is one of letting go that may offer opportunities and possibilities to reconfigure and inter-relate anew with ourselves, others and the world.

This letting-go, or in German 'Gelassenheit', can be translated as serenity, composure or detachment. As such it denotes a non-objectifying ethos of what can be called in paradoxical way an active and ongoing passivity. The ethos of being 'gelassen' involves an accepting attitude by a cultivated and careful 'letting-be'. This letting go is abandoning habitual, representational or appropriating orientations and actions. Such bearing appears to be very challenging in contemporary organizations and economy and leadership with its performance-driven 'practicalism' and corresponding constraints. However, it is exactly because of these increasingly unviable forms of conventional modes of organising and economising all that 'Gelassenheit 'is and will become even more necessary for a more sustainable today and tomorrow to come.

In practicing this 'Gelassenheit', people do not try anymore to manipulate, or master. In its place, and realising a kind of post-heroic mode, they are free for possibilities of eco-scenic be(com)ing to emerge in their own disclosing and vital ways. Significantly, this process is not one of indifference or lack of interest, rather is qualified as an 'engaged letting go' that operates without appropriating projections or totalising closures.

Entering a modus of letting-be, in and through embodied 'ecocening', is realised through a receptive waiting and attending. This more receptive orientation is more an 'active non-doing' in relation to what 'matters', rather than a controlling-driven and fixed willing of conservative forms of 'business as usual'. In particular, letting-be moves from a representational and calculative mode towards more poetic relations. Furthermore, this intermediated via being in the now and, atmospheric sensitivities and forms of dwellings that are a kind of proto-meditative tuning and mindful being.

By cultivating this letting-go those who engage in it silence habitual modes of thinking and doing. By developing this 'Gelassenheit', it may become possible to suspend and redirect instrumental modes and routinised unsustainable behaviour. Therefore it becomes possible to openly obtain encouragements that come from the uplifting depth of other beings in their otherness.

Such orientation is not on a mission to pursue the modernist and adapted eco-modernist anthropocentric project of putting questions to phenomena and forcing them to answer or being exploited or ill-treated or inventing quick fixes in hasty operations.

'Gelassenheit' discourages mindless organising or the exploitive misuse of unsustainable practices or bad leadership in its rigid, intemperate, callous, corrupt, insular [142] or violent modes, including a 'violent innocence' [143].

Embodied and critically engaged letting-go is not the type of depoliticised (secular corporate) mindfulness practices that (re-)produce docile, neoliberal subjects as part of a bio-power(ful) governmentality regime. It is questioning those regimes that, among other issues, hypostatise the tension and dynamism of neoliberal capitalism and fail to cultivate a critical awareness of social, political and ecological factors [144]. Instead it explores and enacts forms of 'commoning', and cultivates a genuine togetherness of "more-than-human commons" [145].

Like mindfulness, engaged 'Gelassenheit' considers the mutually constitutive dimensions of selves to complex social and political ecologies of other subjects and the world as 'inter-jectives', and letting all evolve as embodied in-between also in practically wise organizations [146].

Overall, it is trusted that the concepts and perspectives as drawn here provide possibilities to wisely re-assess and revive the relevance of the embodied dimension of practices in and through organizations as part of the move from the Anthropocene towards an ecocene. Enacting this bodied, performative practice in and beyond organizational life-worlds, pursuit in the spirit of an engaged 'Gelassenheit' may then mediate an organizationally related incarnation and unfoldment of a genuine 'alter-native. An 'alter-native' is here taken literally as 'other-birthly', thus altered economic, political, societal and ethical 'inter-ests' and inter-relationships of sustainable worlds to become are born.

Further critical research on concept and narratives of the Anthropocene, but even more an ecozoic pedagogy [147] and experimental enactments of ways towards an ecocene will be essential. These can provide opportunities to cultivate more prudent, non-hubristic human, especially organizational and leadership practices. Realised and evaluated in everyday life, such research, education and experimentation can then contribute to practically wise sensing, feeling, thinking, deciding, as well as en- and interacting spiralling, cycles of sustainable unfoldment that are as much embodied mindful as 're-evolutionary'.

This all can be envisioned by imagining the realisation as expressed by Paul Goodman (1911–1972): "Suppose you had the revolution you are talking and dreaming about. Suppose your side had won, and you had the kind of society/organization that you wanted. How would you live, you personally, in that society/organization? Start living that way now!"

Funding: "This research received no external funding".

Conflicts of Interest: "The authors declare no conflict of interest."

References

1. Allenby, B.R.; Sarewitz, D.R. *The Techno-human Condition*; MIT Press: Cambridge, MA, USA, 2011.
2. Steffen, W.; Grinevald, J.; Crutzen, P.; McNeill, J. The Anthropocene: Conceptual and Historical Perspectives. *Philos. Trans. R. Soc. A* **2011**, *369*, 842–867. [CrossRef]
3. Steffen, W.; Persson, Å.; Deutsch, L.; Zalasiewicz, J.; Williams, M.; Richardson, K.; Crumley, C.; Crutzen, P.; Folke, C.; Gordon, L.; et al. The Anthropocene: From Global Change to Planetary Stewardship. *AMBIO* **2011**, *40*, 739–761. [CrossRef] [PubMed]
4. Liu, J.; Mooney, H.; Hull, V.; Davis, S.J.; Gaskell, J.; Hertel, T.; Kremen, C. Systems integration for global sustainability. *Science* **2015**, *347*, 1258832. [CrossRef]
5. Serres, M. *Malfeasance: Appropriation Through Pollution?* Stanford University Press: Stanford, CA, USA, 2011.
6. Serres, M. *The Natural Contract*; The University of Michigan Press: Ann Arbor, MI, USA, 1995.
7. Serres, M. *Times of Crisis: What the Financial Crisis Revealed and How to Reinvent Our Lives and Future*; Bloomsbury: London, UK, 2014.
8. Brown, S.D. They have escaped the weight of darkness: The problem space of Michel Serres. In *The Routledge Companion to Reinventing Management Education*; Steyaert, C., Beyes, T., Parker, M., Eds.; Routledge: Basingstoke, UK, 2016.
9. Heikkurinen, P.; Ruuska, T.; Kuokkanen, A.; Russell, S. Leaving Productivism behind: Towards a Holistic and Processual Philosophy of Ecological Management. *Philos. Manag.* **2019**. [CrossRef]
10. Steffen, W.; Broadgate, W.; Deutsch, L.; Gaffney, O.; Ludwig, C. The Trajectory of the Anthropocene: The Great Acceleration, Anthropocene. *Antropocene Rev.* **2015**, *2*, 81–98. [CrossRef]
11. Bendell, J. Deep Adaptation: A Map for Navigating Climate Tragedy IFLAS Occasional Paper 2 www.iflas.info. 2018. Bendell, J. After Climate Despair—One Tale Of What Can Emerge. Jembendell.com. Available online: https://jembendell.wordpress.com/2018/01/14/after-climate-despair-one-tale-ofwhat-can-emerge/despairasagatewaytoourgrowth (accessed on 18 April 2020).
12. Böhm, S.; Bharucha, Z.P.; Pretty, J. Ecocultures: Towards Sustainable Ways of Living. In *Ecocultures: Blueprints for Sustainable Communities*; Böhm, S., Bharucha, Z.P., Pretty, J., Eds.; Routledge: London, UK, 2014.
13. Foster, J.M. Hope after sustainability? tragedy and transformation. *Glob. Discourse* **2017**, *7*, 171–187. [CrossRef]
14. Schimelpfenig, R. The Drama of the Anthropocene: Can Deep Ecology, Romanticism, and Renaissance Science Rebalance Nature and Culture? *Am. J. Econ. Sociol.* **2017**, *76*, 821–1081. [CrossRef]

15. Biermann, F.; Lövbrand, E. Encountering the 'Anthropocene': Setting the scene. In *Anthropocene Encounters: New Directions in Green Political Thinking*; Biermann, F., Lövbrand, E., Eds.; Cambridge University Press: Cambridge, UK, 2019; pp. 1–22.
16. Swyngedouw, E.; Ernsts, H. Interrupting the Anthropo-obScene: Immuno-biopolitics and Depoliticising Ontologies in the Anthropocene. *Theory Cult. Soc.* **2018**, *35*, 3–30. [CrossRef]
17. Parikka, J. *The Anthrobscene*; University of Minnesota Press: Minneapolis, MN, USA, 2015.
18. Zalasiewicz, J. The Extraordinary Strata of the Anthropocene. In *Environmental Humanities: Voices from the Anthropocene*; Oppermann, S., Iovino, S., Eds.; Rowman and Littlefield International: London, UK, 2017.
19. Toadvine, T.A., Jr. Apocalyptic imagination and the silence of the elements. In *Ecopsychology, Phenomenology, and the Environment: The Experience of Nature*; Springer: New York, NY, USA, 2014; pp. 211–221.
20. Morton, T. *Dark Ecology: For a Logic of Future Coexistence*; Columbia University Press: New York, NY, USA, 2016.
21. Morton, T. *Hyperobjects: Philosophy and Ecology after the End of the World*; University of Minnesota Press: Minneapolis, MN, USA, 2013.
22. Neyrat, F. *The Unconstructable Earth An Ecology of Separation*; Fordham: New York, NY, USA, 2018.
23. Küpers, W. Integrating Hope and Wisdom in Organisation in the Anthropocene. In *Organizing Goodness and Hope*; Ericsson, D., Kostera, M., Eds.; Edward Elgar: Cheltenham, UK, 2019; pp. 72–84.
24. Blühdorn, I. Post-capitalism, post-growth, post-consumerism? Eco-political hopes beyond sustainability. *Glob. Discourse* **2017**, *7*, 42–61. [CrossRef]
25. Benson, M.; Craig, R.K. The End of Sustainability. *Soc. Nat. Resour.* **2014**, *27*, 777–782. [CrossRef]
26. Foster, J.M. *After Sustainability*; Earthscan: Abingdon, UK, 2015.
27. Kagan, S. *Art and Sustainability: Connecting Patterns for a Culture of Complexity*; Transcript: Bielefeld, Germany, 2013.
28. Kagan, S. Artful Sustainability: Queer-convivialist life-art and the artistic turn in sustainability research. *Transdiscipl. J. Eng. Sci.* **2017**, *8*, 151–168. [CrossRef]
29. Bonneuil, C.; Fressoz, J.-B. *The Shock of the Anthropocene: The Earth, History, and Us*; Verso Books: New York, NY, USA, 2016.
30. Sloterdijk, P. Das Anthropozän: Ein Prozess-Zustand am Rande der Erd-Geschichte? In *Das Anthropozän: Zum Stand der Dinge*; Renn, J., Scherer, B., Eds.; Matthes & Seitz: Berlin, Germany, 2015.
31. Slaughter, R.A. *Futures beyond Dystopia: Creating Social Foresight*; Routledge: London, UK, 2004.
32. Campbell, N.; McHugh, G.; Ennis, O. Climate Change is Not a Problem: Speculative Realism at the End of Organization. *Organ. Stud.* **2019**, *40*, 725–744. [CrossRef]
33. Zalasiewicz, J.; Williams, M.; Steffen, W.; Crutzen, P. The new world of the Anthropocene Environ. *Sci. Technol.* **2010**, *44*, 2228–2231. [CrossRef] [PubMed]
34. Arias-Maldonado, M. The Anthropocenic Turn: Theorizing Sustainability in a Postnatural Age. *Sustainability* **2016**, *8*, 1–17. [CrossRef]
35. Barry, J.; Mol, A.; Zito, A. Climate Change Ethics, Rights, and Policies: An Introduction. *Environ. Politics* **2013**, *22*, 361–376. [CrossRef]
36. Agamben, G. *The Open Man and Animal*; Stanford University Press: Redwood City, CA, USA, 2009.
37. Malm, A.; Hornborg, A. The geology of mankind? A critique of the Anthropocene narrative. *Anthr. Rev.* **2014**, *1*, 62–69. [CrossRef]
38. Baskin, J. Paradigm dressed as Epoch: The Ideology of the Anthropocene. *Environ. Values* **2015**, *24*, 9–29. [CrossRef]
39. Hamilton, C. *Earthmasters: Playing God with the Climate*; Allen & Unwin: Crows Nest, NSW, Australia, 2013.
40. Hamilton, C. *Defiant Earth: The Fate of Humans in the Anthropocene*; Polity Press: Cambridge, UK, 2017.
41. Neyrat, F. *La part inconstructible de la Terre. Critique du géo-constructivisme*; Seuil: Paris, France, 2016.
42. Howe, C. Anthropocenic Ecoauthority: The Winds of Oaxaca Anthropological Quarterly, Energopower and Biopower in Transition. *Anthropol. Q.* **2013**, *87*, 381–404. [CrossRef]
43. Arias-Maldonado, M. *Environment & Society: Socionatural Relations in the Anthropocene*; Springer VS: Wiesbaden, Germany, 2015.
44. Haraway, D. *The Companion Species Manifesto: Dogs, People, and Significant Otherness*; Prickly Paradigm Press: Chicago, IL, USA, 2003.
45. Küpers, W. To be physical is to 'inter-be-come'. Beyond empiricism and idealism towards embodied leadership that matters". In *'Physicality of Leadership', Gesture, Entanglement, Taboo, Possibilities*; Ladkin, D., Taylor, S., Eds.; Emerald: London, UK, 2014; pp. 83–108.

46. Küpers, W. Embodied inter-practices of leadership, special issue on 'The Materiality of Leadership: Corporeality and subjectivity'. *Leadership* **2013**, *9*, 335–357. [CrossRef]
47. Küpers, W. Embodied performance and performativity in organizations and management", Special Issue: 'Putting critical performativity to work'. *Management* **2017**, *20*, 89–106.
48. Berkhout, F. Anthropocene Futures. *Anthr. Rev.* **2014**, *1*, 154–159. [CrossRef]
49. Pyyhtinen, O.; Tamminen, S. We have never been only human: Foucault and Latour on the question of the anthropos. *Anthropol. Theory* **2011**, *11*, 135–152. [CrossRef]
50. Zalasiewicz, J.; Williams, M.; Fortey, R.; Smith, A.; Barry, T.L.; Coe, A.L.; Bown, P.R.; Rawson, P.F.; Gale, A.; Gibbard, P.; et al. Stratigraphy of the Anthropocene. *Philos. Trans. R. Soc. A* **2011**, *369*, 1036–1055. [CrossRef] [PubMed]
51. Steffen, W.; Crutzen, P.J.; McNeill, J.R. The Anthropocene: Are humans now overwhelming the great forces of Nature? *Ambio* **2011**, *36*, 614–621. [CrossRef]
52. Mauelshagen, F. Bridging the Great Divide. The Anthropocene as a Challenge to the Social Sciences and Humanities. In *Religion and the Anthropocene*; Deane-Drummond, C., Bergmann, S., Vogt, M., Eds.; University of Oregon Press: Eugene, OR, USA, 2017.
53. Marris, E.; Garden, R. *Saving Nature in a Post-Wild World*; Bloomsbury: New York, NY, USA, 2011.
54. Horn, E.; Jenseits der Kindeskinder. Nachhaltigkeit im Anthropozän. *Merkur* **2017**. Available online: https://www.merkur-zeitschrift.de/2017/02/23/jenseits-der-kindeskinder-nachhaltigkeit-im-anthropozaen/ (accessed on 18 April 2020).
55. Kalonaityte, V. When rivers go to court: The Anthropocene in organization studies through the lens of Jacques Rancière. *Organization* **2018**, *25*, 517–532. [CrossRef]
56. Thompson, A. Anthropocentrism: Humanity as Peril and Promise. In *The Oxford Handbook of Environmental Ethics*; Gardiner, S.M., Thompson, A., Eds.; Oxford University Press: Oxford, UK, 2019.
57. Callicott, J. *In Defense of the Land Ethic*; SUNY Press: Albany, NY, USA, 1989.
58. Rolston, H. *Conserving Natural Value*; Columbia University Press: New York, NY, USA, 1994.
59. Rolston, H. The Anthropocene! Beyond the Natural? In *Stephen M. Gardiner and Allen Thompson, The Oxford Handbook of Environmental Ethics*; Oxford University Press: Oxford, UK, 2019.
60. Toadvine, T. Phenomenology and Environmental Ethics. In *The Oxford Handbook of Environmental Ethics*; Gardiner, S.M., Thompson, A., Eds.; Oxford University Press: Oxford, UK, 2019; pp. 222–245.
61. Küpers, W. Embodied Inter-Be(com)ing – The contribution of Merleau-Ponty's relational Ontology for a processual understanding of Chiasmic Organising. In *Oxford Handbook of Process Philosophy and Organization Studies*; Helin, J., Hernes, T., Hjorth, D., Holt, R., Eds.; Oxford University Press: Oxford, UK, 2014; pp. 413–431.
62. Merleau-Ponty, M. *Phenomenology of Perception*; Translated by Landes, D.A.; Routledge: London, UK; New York, NY, USA, 2012.
63. Merleau-Ponty, M. *The Visible and the Invisible*; Northwestern University Press: Evanston, IL, USA, 1995.
64. Merleau-Ponty, M. *Nature. Course Notes from the College de France*; Northwestern University Press: Evanston, IL, USA, 2003.
65. Cecil, P. Changing Cultures in Organizations: A Process of Organicization. *Concrescence* **2004**, *5*, 1445–4297.
66. Toadvine, T. *Merleau-Ponty's Philosophy of Nature*; Northwestern University Press: Evanston, IL, USA, 2009.
67. Cataldi, S.; Hamrick, W. *Merleau-Ponty and Environmental Philosophy: Dwelling on the Landscapes of Thought*; State University of New York Press: Albany, NY, USA, 2007.
68. Kleinberg-Levin, D. *Before the voice of reason. Echoes of Responsibility in Merleau-Ponty's Ecology and Levinas's Ethics*; State University of New York Press: Albany, NY, USA, 2008.
69. Brown, C.; Toadvine, T. *Eco-Phenomenology: Back to the Earth Itself*; State University of New York Press: Albany, NY, USA, 2003.
70. Mazis, G.A. *Earthbodies, Rediscovering our Planetary Senses*; State University of New York Press: Albany, NY, USA, 2002.
71. Mazis, G.A. *Merleau-Ponty and the Face of the World: Silence, Ethics, Imagination and Poetic Ontology*; State University of New York Press: Albany, NY, USA, 2016.
72. Booth, R. Merleau-Ponty, Correlationism, and Alterity. *PhænEx* **2018**, *12*, 37–58. [CrossRef]
73. Iovino, S.; Oppermann, S. *Material Ecocriticism*; Indiana University Press: Bloomington, IN, USA, 2014.
74. Swimme, B.; Berry, T. *The Universe Story From the Primordial Flaring Forth to the Ecozoic Era*; Harper: San Francisco, CA, USA, 1992.

75. Korchagina, N. Organizing for an ecologically sustainable world: Reclaiming nature as wonder through Merleau-Ponty. *Epherema* **2019**, *18*, 805–815.
76. Küpers, W. Embodied, Relational Practices of Human and Non-Human in a Material, Social, and Cultural Nexus of Organizations. *Cult. Open J. Study Cult.* **2016**, *2*. Available online: http://geb.uni-giessen.de/geb/volltexte/2016/12352/ (accessed on 18 April 2020).
77. Latour, B. *Facing Gaia: A New Enquiry into Natural Religion, The Gifford Lectures*; University of Edinburgh: Edinburgh, UK, 2013; Available online: http://www.bruno-latour.fr/node/486 (accessed on 30 April 2019).
78. Buchli, V. Introduction. In *The material Culture Reader*; Buchli, V., Ed.; Oxford and Berg: New York, NY, USA, 2002.
79. Swyngedouw, E. The non-political politics of climate change. *ACME* **2013**, *12*, 1–8.
80. Küpers, W. Embodied Responsive Ethical Practice: The Contribution of Merleau-Ponty for a Corporeal Ethics in Organisations". *Electron. J. Bus. Ethics Organ. Stud. (EJBO)* **2015**, *20*, 30–45.
81. Neimanis, A.; Suzi, H.; Cecilia, Å. Feminist Posthumanist Imaginaries of Climate Change. In *Research Handbook on Climate Governance*; Bäckstrand, K., Lövbrand, E., Eds.; Edward Elgar Publishing: Northampton, CA, USA, 2015; Volume 1, pp. 480–490.
82. Fesmire, S. Ecological Imagination in Moral Education, East and West. *Contemp. Pragmatism* **2012**, *9*, 205–222. [CrossRef]
83. Garrard, G.; Handwerk, G.; Wilke, S. Introduction: Imagining Anew: Challenges of Representing the Anthropocene. *Environ. Humanit.* **2014**, *5*, 149–153. [CrossRef]
84. Gosling, J.; Case, P. Social dreaming and ecocentric ethics: Sources of non-rational insight in the face of climate change catastrophe. *Organization* **2013**, *20*, 705–721. [CrossRef]
85. Werhane, P.H. Moral imagination and systems thinking. *J. Bus. Ethics* **2002**, *38*, 33–42. [CrossRef]
86. Werhane, P.H. Mental models, moral imagination and systems thinking in the age of globalization. *J. Bus. Ethics* **2008**, *78*, 463–474. [CrossRef]
87. Narvaez, D.; Mrkva, K. Creative moral imagination. In *The Ethics of Creativity*; Moran, S., Cropley, D., Kaufman, J., Eds.; Palgrave MacMillan: New York, NY, USA, 2014; pp. 25–45.
88. Hargrave, T.J. Moral imagination, collective action, and the achievement of moral outcomes. *Bus. Ethics Q.* **2009**, *19*, 85–102. [CrossRef]
89. Bachelard, G. *The Poetics of Reverie*; Beacon Press: Boston, MA, USA, 1971.
90. Smith, J. New Bachelards? Reveries, Elements and Twenty-First Century Materialism. *Alter Mod.* **2012**, 156–167. [CrossRef]
91. Ghosh, A. *The Great Derangement. Climate Change and the Unthinkable*; University of Chicago Press: Chicago, IL, USA, 2016.
92. Weber, A. *Enlivenment: Towards a Fundamental Shift in the Concepts of Nature, Culture and Politics*; Heinrich–Böll–Stiftun: Berlin, Germany, 2013.
93. Weber, A. *Enlivenment: Toward a Poetics for the Anthropocene*; MIT Press: Cambridge, MA, USA, 2019.
94. Braidotti, R. *Transpositions: On Nomadic Ethics*; Polit: Cambridge, UK; Malden, MA, USA, 2006.
95. Segall, M. Logos of a Living Earth: Towards a Gaian Praxecology. 2009. Available online: https://footnotes2plato.com/2009/11/21/logos-of-a-living-earth-towards-a-gaian-praxecology/ (accessed on 18 April 2020).
96. Xiang, W.-N. *Ecophronesis: The Ecological Practical Wisdom for and from Ecological Practice, In Ecological Wisdom Theory and Practice*; Yang, B., Young, R.F., Eds.; Springer Nature: Singapore, 2019.
97. Scharper, S. *Redeeming the Time: A Political Theology of the Environment*; Continuum: New York, NY, USA, 1997.
98. Scharper, S. *For Earth's Sake: Toward a Compassionate Ecology*; Novalis: Toronto, ON, USA, 2013.
99. Hathaway, M. The Practical Wisdom of Permaculture: An Anthropoharmonic Phronesis for Moving towards an Ecological Epoch. *Environ. Ethics* **2015**, *37*, 445–463. [CrossRef]
100. Rosa, H.; Dörre, K.; Lessenich, S. Appropriation, activation and acceleration: The escalatory logics of capitalist modernity and the crises of dynamic stabilization. *Theory Cult. Soc.* **2017**, *34*, 53–73. [CrossRef]
101. Rosa, H. *Social Acceleration: A new Theory of Modernity*; Columbia: New York, NY, USA, 2013.
102. Rosa, H. *Resonance: A Sociology of Our Relationship to the World*; Polity: Cambridge, UK, 2019.

103. Küpers, W. "Embodied Inter-Practices in Resonance as New Forms of Working in Organisations". In *Experiencing the New World of Work*; Aroles, J., Dale, K., de Vaujany, F., Eds.; Cambridge University Press: Cambridge, UK, 2020; (forthcoming).
104. Totten, M. Flourishing Sustainably in the Anthropocene? Known Possibilities and Unknown Probabilities. In *Reference Module in Earth Systems and Environmental Sciences*; Elsevier: New York, NY, USA, 2018.
105. Seray, E.; Calás, M.; Smircich, L. Ecologies of Sustainable Concerns: Organization Theorizing for the Anthropocene. *Gend. Work Organ.* **2018**, *25*, 222–245.
106. Wright, C.C.; Nyberg, D.; Richards, L.; Freund, J. Organizing in the Anthropocene. *Organization* **2018**, *25*, 455–471. [CrossRef]
107. Shrivastava, P. Castrated environment: Greening organizational studies. *Organ. Stud.* **1994**, *15*, 705–726. [CrossRef]
108. Purser, R.E.; Park, C.; Montuori, A. Limits to anthropocentrism: Toward an ecocentric organization paradigm? *Acad. Manag. Rev.* **1995**, *20*, 1053–1089. [CrossRef]
109. Foster, J.B.; Jermier, J.M.; Shrivastava, P. Global environmental crisis and ecosocial reflection and inquiry: Introduction to Organization & Environment. *Organ. Environ.* **1997**, *10*, 5–11.
110. Linnenluecke, M.K.; Griffiths, A. *The Climate Resilient Organization: Adaptation and Resilience to Climate Change and Weather Extremes*; Edward Elgar: Cheltenham, UK, 2015.
111. Heikkurinen, P.; Rinkinen, J.; Järvensivu, T.; Wilén, K.; Ruuska, T. Organising in the Anthropocene: An ontological outline for ecocentric theorising. *J. Clean. Prod.* **2016**, *113*, 705–714. [CrossRef]
112. Bell, V. Declining performativity: Butler, Whitehead and ecologies of concern. *Theory Cult. Soc.* **2012**, *29*, 107–123.
113. Küpers, W. Phenomenology of embodied and artful design for creative and sustainable inter-practicing in organisations. *J. Clean. Prod.* **2016**, *135*, 1436–1445.
114. Küpers, W. *Phenomenology of the Embodied Organization—The contribution of Merleau-Ponty for organisation studies and practice*; Palgrave: London, UK, 2015.
115. Ciocirlan, C.E. Environmental workplace behaviors: Definition matters. *Organ. Environ.* **2017**, *30*, 51–70.
116. Mathews, F. Towards a deeper philosophy of biomimicry. *Organ. Environ.* **2011**, *24*, 364–387. [CrossRef]
117. Jones, D.R. The 'Biophilic Organization': An Integrative Metaphor for Corporate Sustainability. *J. Bus. Ethics* **2016**, *138*, 401–416. [CrossRef]
118. Driscoll, C.; Starik, M. The primordial stakeholder: Advancing the conceptual consideration of stakeholder status for the natural environment. *J. Bus. Ethics* **2004**, *49*, 55–73. [CrossRef]
119. Haigh, N.; Griffiths, A. The natural environment as a primary stakeholder: The case of climate change. *Bus. Strategy Environ.* **2009**, *18*, 347–359. [CrossRef]
120. Waddock, S. We are all stakeholders of Gaia: A normative perspective on stakeholder thinking. *Organ. Environ.* **2011**, *24*, 192–212. [CrossRef]
121. Küpers, W. Integral pheno-practice of wisdom in management and organization. *Soc. Epistemol.* **2007**, *22*, 169–193. [CrossRef]
122. Shove, E.; Spurling, N. (Eds.) *Sustainable Practices—Social Theory and Climate Change*; Routledge: London, UK, 2013.
123. Michael, J.; Littig, B. Sustainable Practices. In *International Encyclopedia of the Social & Behavioral Sciences*, 2nd ed.; Wright, J.D., Ed.; Elsevier: Oxford, UK, 2015; pp. 834–838.
124. Hoffman, A.J.; Bazerman, M.H. Changing Practice on Sustainability: Understanding and Overcoming the Organizational and Psychological Barriers to Action. In *Organizations and the Sustainability Mosaic: Crafting Long-Term Ecological and Societal Solutions*; Sharma, S., Starik, M., Husted, B., Eds.; Edward Elgar: Northampton, MA, USA, 2007; pp. 84–105.
125. Van Kleef, G.A.; Oveis, C.; van der Löwe, I.; LouKogan, A.; Goetz, J.; Keltner, D. Power, distress, and compassion: Turning a blind eye to the suffering of others. *Psychol. Sci.* **2008**, *19*, 1315–1322. [CrossRef]
126. Rancière, J. *Disagreement: Politics and Philosophy*; UMP: Minneapolis, MN, USA, 1998.
127. Rancière, J. *The politics of aesthetics. The Distribution of the Sensible*; Continuum: London, UK, 2004.
128. Rancière, J. *The Aesthetic Unconscious Aesthetics and Its Discontents*; Polity Press: Cambridge, UK, 2009.
129. Rancière, J. *Dissensus: On Politics and Aesthetics*; Polity Press: Cambridge, UK, 2010.
130. Vihalem, M. Everyday aesthetics and Jacques Rancière:r econfiguring the common field of aesthetics and politics. *Journal of Aesthetics Culture* **2018**, *10*, 1506209. [CrossRef]

131. Beyes, T. Reframing the Possible: Rancierian aesthetics and the study of organization. *Aesthesis Int. J. Art Aesthet. Manag. Organ. Life* **2008**, *2*, 32–41.
132. Deranty, J.-P. Rancière and Contemporary Political Ontology. *Theory Event* **2003**, 6. [CrossRef]
133. Birkeland, I.; Burton, R. Cultural Sustainability and the Nature-Culture Interface. In *Cultural Sustainability and the Nature-Culture Interface*; Parra, C., Siivonen, K., Eds.; Routledge: Abingdon, UK, 2018.
134. Warren, S. Empirical challenges in organizational aesthetics research: Towards a sensual methodology. *Organ. Stud.* **2008**, *19*, 559–580. [CrossRef]
135. Sandelands, L.E.; Srivatsan, V. The Problem of Experience in the Study of Organizations. *Organ. Stud.* **1993**, *14*, 1–22. [CrossRef]
136. Park Lala, A.; Kinsella, E.A. Embodiment in research practices: The body in qualitative research. In *Creative Spaces for Qualitative Researching: Living Research*; Higgs, J., Titchen, A., Horsfall, D., Bridges, D., Eds.; Sensem: Rotterdam, The Netherlands, 2011; pp. 77–86.
137. Thanem, T.; King, D. *Embodied Research Methods*; Sage: London, UK, 2019.
138. Delanty, G.; Mota, A. Governing the anthropocene: Agency, governance, knowledge. *Eur. J. Soc. Theory* **2017**, *20*, 9–38. [CrossRef]
139. Mickey, S. Elemental Imagination: Deconstructive Phenomenology and the Sense of Environmental Ethics. In *Ecopsychology, Phenomenology, and the Environment: The Experience of Nature*; Vakoch, D.A., Castrillón, F., Eds.; Springer: New York, NY, US, 2014; pp. 159–176.
140. Hannis, M. *Virtue is Good for You: The Politics of Ecological Eudaimonism, Ecological Virtues*; Chang, D., Bai, H., Eds.; University of Regina Press: Saskatchewan, Canada, 2017.
141. Heikkurinen, P. Degrowth by Means of Technology? A Treatise for an Ethos of Releasement. *J. Clean. Prod.* **2016**, *197*, 1654–1665. [CrossRef]
142. Kellerman, B. *Bad Leadership What It Is, How It Happens, Why It Matters*; Harvard Business School Press: Boston, MA, USA, 2004.
143. Vince, R.; Mazen, A. Violent innocence: A contradiction at the heart of leadership. *Organ. Stud.* **2014**, *35*, 189–207. [CrossRef]
144. Walsh, Z. Mindfulness under neoliberal governmentality: Critiquing the operation of biopower in corporate mindfulness and constructing queer alternatives. *J. Manag. Spiritual. Relig.* **2018**, *15*, 109–122. [CrossRef]
145. Walsh, Z. Contemplating the More-than-Human Commons. *Arrow A J. Wakeful Soc. Cult. Politics* **2018**, *5*, 5–18. Available online: https://arrow-journal.org/contemplating-the-more-than-human-commons/#fn-2044-9 (accessed on 18 March 2020).
146. Küpers, W. The embodied Inter-Be(com)ing of Spirituality. The in-between as spiritual sphere in practically wise organizations". In *Managing VUCA by an Integrative Self-Managed way: Enhancing Integrating Simplification Theory*; Nandram, S., Ed.; Springer: London, UK, 2017; pp. 247.
147. Vargas Roncancio, I.; Temper, L.; Sterlin, J.; Smolyar, N.L.; Sellers, S.; Moore, M.; Melgar-Melgar, R.; Larson, J.; Horner, C.; Erickson, J.D.; et al. From the Anthropocene to Mutual Thriving: An Agenda for Higher Education in the Ecozoic. *Sustainability* **2019**, *11*, 3312. [CrossRef]

MDPI
St. Alban-Anlage 66
4052 Basel
Switzerland
Tel. +41 61 683 77 34
Fax +41 61 302 89 18
www.mdpi.com

Sustainability Editorial Office
E-mail: sustainability@mdpi.com
www.mdpi.com/journal/sustainability

www.ingramcontent.com/pod-product-compliance
Lightning Source LLC
LaVergne TN
LVHW070616170726
843515LV00004B/92
* 9 7 8 3 0 3 9 3 6 9 5 6 0 *